Generalized Optomechanics and Its Applications

Quantum Optical Properties of
Generalized Optomechanical System

Generalized Optomechanics and Its Applications

Quantum Optical Properties of Generalized Optomechanical System

Jin-Jin Li

Shanghai Jiao Tong University, China

Ka-Di Zhu

Minhang Shanghai Jiao Tong University, China

World Scientific

NEW JERSEY · LONDON · SINGAPORE · BEIJING · SHANGHAI · HONG KONG · TAIPEI · CHENNAI

Published by

World Scientific Publishing Co. Pte. Ltd.

5 Toh Tuck Link, Singapore 596224

USA office: 27 Warren Street, Suite 401-402, Hackensack, NJ 07601

UK office: 57 Shelton Street, Covent Garden, London WC2H 9HE

British Library Cataloguing-in-Publication Data
A catalogue record for this book is available from the British Library.

ISBN 978-981-4417-03-7

Printed by FuIsland Offset Printing (S) Pte Ltd Singapore

Preface

People always tend to pursuit the best manipulation of the light speed, e.g., slow light, fast light or even complete stopped light. To achieve this goal, some advanced technologies like electromagnetically induced transparency (EIT) or coherent population oscillation (CPO) have been put forward to manipulate the light speed to a satisfied magnitude as much as possible. Recently, a hot frontier topic — "optomechanics" has emerged to achieve light speed control in mechanical systems, which combine optics and mechanics in microscopic materials.

A typical cavity optomechanical system consists of a mechanical oscillator coupled to the optical field in a cavity. The control of light propagation based on these typical optomechanical systems has been studied by many researchers experimentally and theoretically. These works provide a particularly helpful toolbox for generating and controlling new quantum optical properties of mechanical-like systems, such as coupled nanomechanical systems, carbon nanotubes, surface plasmons, Bose-Einstein condensate (BEC), and other condensed matter nanomaterials like quantum dots and graphene. These mechanical-like structures, with specific characteristics, have some commons with cavity optomechanical systems, and can develop a generalized optomechanical system (GOS). In this book, we extend the concept of "optomechanics" to "generalized optomechanics", and provide a balanced introduction of GOS by discussing its structures and light propagation properties in this rapidly developing field. We also outline some exciting emerging applications, e.g., generation of slow light and fast light, nonlinear Kerr modulator, quantum optical transistor, quantum memory, optical mass sensing and other ultrasensitive detections.

It seems clear that the development of a new field of generalized optomechanics offers a platform for further quantum phenomena which involves both photons and mechanical systems. This will soon make optics and condensed matter physics merge together well.

Keywords: Optomechanics, Optomechanical system, Light propagation, Quantum dot, Nanomechanical resonator, Surface plasmon, Carbon nanotube, Superconducting microwave cavity, Bose-Einstein condensate, DNA molecule, Quantum optical transistor, Optical mass sensing.

Jin-Jin Li and Ka-Di Zhu

Acknowledgments

This work was supported by the National Natural Science Foundation of China (No. 10774101, No. 10974133 and No. 11274230), the National Ministry of Education Program for Ph.D, the Scholarship Award for Excellent Doctoral Student granted by Ministry of Education, China, and the Academic Award for Distinguished Doctoral Candidates granted by SJTU.

We would like to express our gratitude to many people who provided valuable help. They are Prof. Chun-Fang Li, Prof. Zhi-Ming Zhang, Dr. Wei He, Dr. Bin Chen, Dr. Cheng Jiang, Dr. Huan Wang and Dr. Zhi-En Lu. We would also like to thank Don Mak for enabling us to publish this book.

Contents

Chapter 1

Introduction

1.1 Optomechanical systems

Whether in classical or in quantum physics, the harmonic oscillator is the simplest known example of a study. Its mechanical performance, namely, mechanical oscillator, is probably the most concrete and intuitive implementation subject [Vahala (2008)]. In the past few decades, the mechanical oscillator has been investigated with different science and technologies. Its applications range from atomic force microscope cantilever for micro size to nano-mechanical oscillator mirror interferometer for the detection of gravitational waves, displacement etc. [Vahala (2005); Gigan (2006); Schliesser (2006); Ward (2009)]. The quantum control of the mechanical motion is state of the art, which spans the size range from hundreds of nanometers in the case of nano-electro-mechanical or nano-opto-mechanical systems (NEMS/NOMS) to tens of centimeters in the case of gravitational wave antennae [Xuereb (2010); Kiesel (2010)].

In recent years, the combination of the optical cavity and mechanical systems has given rise to a rapid development of the research field — cavity optomechanics, which confines light to small volumes by resonant recirculation, composed of a driven high-frequency optical cavity and a high-Q, low-frequency mechanical resonator [Kiesel (2010); Marquardt (2009)]. The purpose of cavity optomechanical system is to investigate the interaction of light with a mechanical oscillator between the two end mirrors via radiation pressure (one is fixed mirror, the other is movable mirror), where the strength of the interaction is enhanced by using an optical cavity (Fig.1.1(a)). The cavity free spectral range $c/2L$ (c is the speed of light in the vacuum and L is the effective cavity length [Vahala (2005); Schliesser (2006)]) is much larger than the frequency of movable mirror (ω_m). To

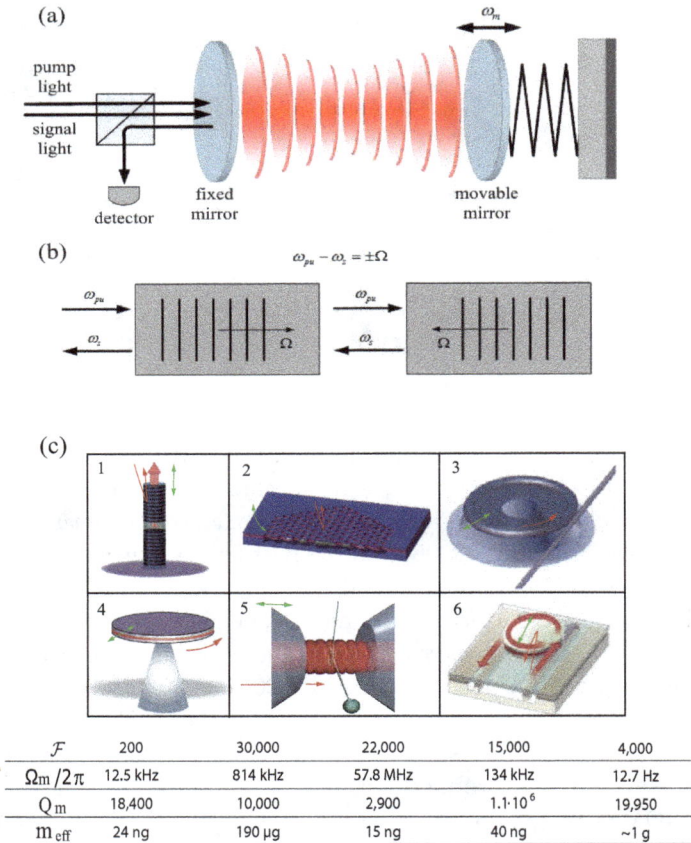

Fig. 1.1 (a) Schematic diagram of a Fabry-Perot cavity with a movable mirror in the presence of a strong driving field and a weak probe field. (b) Schematic diagram of the SBS process in a real material system. ω_{pu} and ω_{pr} are the frequencies of pump field and probe field, respectively, and Ω is the frequency of acoustic wave in the material system. (c) Different physical realizations of cavity optomechanical experiments employing Fabry-Perot cavity, ultracold atoms mediate cavity, microcavity add/drop filter with waveguides, photonic crystal and whispering gallery modes [Vahala (2008, 2003)]. Red and green arrows represent the optical trajectory and mechanical motion, respectively.

account for the mirror suspension or the internal mechanical vibration modes of a mirror, it is assumed that the end mirror is free to oscillate. The high-reflective end mirror enhances the number of roundtrips photons between the beginning mirror and the end mirror and such effect is very sensitive to the position of the end mirror.

The incident optical field, resonating with the cavity mode, produces a lot of circulating energy in the cavity. In turn, the circulating energy can create pressure on the cavity mirrors, causing the movable mirror to move. Reciprocally, the mirror motion results in a new optical round trip condition, which modifies the detuning of the cavity resonance with respect to the incident field. This will cause the system to simply establish a new, static equilibrium condition, and produce a wealth of interesting effects and give rise to a host of long-anticipated phenomena in optomechanical systems.

In turn, the change of cavity's length alters the distribution of circulating intensity. This variation acts as an all-optical diffraction grating in the cavity field moving back and forth with the mechanical resonator frequency ω_m. While the probe light travels in the cavity, the mutual interaction between input lights and the grating leads to the scattering of photons. If the probe light moves in the same direction as the diffraction grating, pump photons will be scattered into probe light, and hence a Stokes process occurs. On the contrary, if they move in different directions, probe light will be scattered into pump field, which results in an anti-Stokes process. Such behavior is very similar to the Stimulated Brillouin Scattering (SBS) in real material systems, in which an acoustic wave of frequency Ω is produced by the mutual interaction between light fields and material system. Through the process of electrostriction, the material system responses to the input fields by the fluctuations of dielectric constant which acts as a moving diffraction grating with frequency Ω as shown in Fig.1.1(b) [Boyd (2008); Gauthier (2009)].

Figure 1.1(c) shows the different physical realization of cavity optomechanical experiments, while implementing in employing 3D-confined cavity, cantilevers, micro-mirrors, micro-cavities, nano-membranes and macroscopic mirror modes [Vahala (2008)]. Devices based on optomechanical systems are already indispensable for a wide range of applications and studies, such as displacement measurement, cavity sensors, optical tweezer, gravity measurement etc. Figure 1.1(c)-1 shows a single-photon source, where a single quantum dot is embedded in the core of a micropost (or micropillar) 3D cavity. The spontaneously emitting of a photon produced by the core quantum dot occurs via the Purcell effect through the cavity top [Solomon (2000)]. Figure 1.1(c)-2 shows a photonic crystal defect microcavity laser. The microcavity is formed by dry etching a hexagonal array of holes and subsequent selective etch of an interior region, creating a thin membrane. One hole is left unetched creating a defect in the array and therefore a defect mode in the optical spectrum [Painter (1999)]. Figure 1.1(c)-3 shows an optical field which is coupled to a fibre-taper waveguide and subsequently is guided within and along the periphery of the microtoroid in a whispering gallery mode [Armani (2003)]. Figure 1.1(c)-4 and 6 are the optomechanical systems in microdisk whispering gallery [Gayral (1999)] and semiconductor polymer add/drop filter [Djordjev (2006); Rabiei (2006)], respectively. In Fig.1.1(c)-5, an ultracold atom is entrained in an orbital motion before escaping. The optical transmission probing of the cavity serves as an ultra-sensitive measurement of atomic location, due to the coupling energy which depends on the amplitude of the vacuum cavity field near the atom.

1.2   Previous research

For cavity optomechanics, the pioneering works have always focused on the measurement and control of the mechanical motion while coupling with external optical fields via radiation pressure, like mechanical damping detection [Vahala (2008)], position measurement [Schliesser (2006)] and the cooling of motion [Zwickl (2008)], which covering a huge range of scales from macroscopic mirrors in the laser interferometer gravitational wave observatory to nano- or micro-mechanical cantilevers, vibrating micro-toroids, and membranes.

     The laser acting back on the mechanical motion is perturbed by the mirror, which gives rise to the dynamical backaction. In the cavity, this backaction leads to the parametric amplification, the negative effective damping and the phase and amplitude modulation of the optical amplitude. The sidebands displaced from the optical cavity frequency $\pm\omega_m$ come from this modulation. The lower and upper sidebands are referred as Stokes and anti-Stokes signals, respectively. The anti-Stokes upper sideband comes from a process that removes energy $\hbar\omega_m$ from the mechanical oscillator, while the lower sideband comes from a cavity photon which loses energy $\hbar\omega_m$ by creating a phonon inside the mechanical oscillator. The intensity difference $2\hbar\omega_m$ between the Stokes and anti-Stokes sidebands declares the energy transfer between mechanical motion and the optical fields, which needs the mechanical cooling down to the ground state as much as possible. The frequently approached optomechanical cooling consists of Doppler-cooling in Bragg mirrors [Marquardt (2009); Karrai (2008)] and "Active feedback cooling" [Kleckner (2006); Vitali (2008)].

The first mechanical damping detection was achieved by Braginsky *et al.* [Braginsky (1967, 1970)] based on the radiation produced by an excited oscillator, using quantum optical method. Since 1990, they have investigated the quantum effects inside an optical cavity through radiation pressure, such as the generation of squeezed light via the optomechanical Kerr effect, the quantum non-demolition measurements of photon number, and the mechanical quantum noise reduction. Recently, Okamoto and colleagues demonstrated vibration amplification, damping, and self-oscillations induced by carrier excitation in optomechanical micro-resonators scheme. They declare that the optomechanical coupling does not require any optical cavities but instead relies on the piezoelectric effect generated by photoinduced carriers [Okamoto (2011)].

For mechanical cooling, until now, experiments have not yet get the ground state cooling, though Aspelmeyer and Kippenberg *et al.* have recently reported a laser cooling of micro-optomechanical resonator that reaches to a level of 30 thermal quanta [Aspelmeyer (2009); Kippenberg (2009)]. In order to get larger mechanical quality factor and to avoid external heating caused by light absorption, the current challenges of mechanical cooling mechanical to the ground state is the cryogenic operation which comes from a low bulk temperature.

1.3 Recent development

As discussed above, the past study mainly focused on the measurement and control of the quantum states and dynamics of optomechanical oscillator while driving lasers, e.g., quantum cooling to the ground state, nonlinear instability and amplification, displacement readout, nonclassical squeezing and entanglement [Marquardt (2009); Carmon (2005); Metzger (2008); Marquardt (2006)].

However, the mechanical oscillator can also affect the dynamics of cavity field and give rise to plenty of unexpected magic optical phenomena. Some theoretical works are in the forefront of experiments and put forward a series of optical effects and light propagation devices, such as slow light and fast light, vacuum Rabi frequency measurement, nonlinear optical modulator, quantum optical sensors and transistors, etc. With the availability of high-quality optomechanical devices, these early theoretical researches have now become an important basis for a whole new field of quantum optomechanics that aims at exploiting the quantum regime of mechanical resonators and the light propagation by means of quantum optics.

In 2009, Schliesser *et al.* first proposed the analog of electromagnetically induced transparency(EIT) in optomechanical systems in his thesis, where they referred to this effect as "optomechanically induced transparency (OMIT)" [Schliesser (2009)]. Since then, EIT effects in these coupled systems are studied both theoretically and experimentally. In 2010, Agarwal *et al.* theoretically investigated the light propagation in cavity optomechanical system [Agarwal (2010)]. They demonstrated the existence of EIT in the output field of a cavity optomechanical system under the action of a pump laser and a probe laser. Zhu *et al.* [Zhu and He (2010a); Zhu and Li (2010b)] studied all-optical control of light group velocity with a cavity optomechanical system and the radiation pressure induced normal mode splitting in this optomechanical system, and further predicted the existence of slow and fast light simultaneously in this system only simply by changing the pump-cavity detuning. Other theoretical works involved the studies of optomechanical displacement transducer [Dobrindt (2010); Stannigel (2010)], slow light and stopped light in an optomechanical crystal array [Painter (2011)], multistability of EIT in atom-assisted optomechanical cavities [Nori (2011)] and cavity quantum optomechanics in ultracold atoms [Bhattacherjee (2009)].

Furthermore, recent experiments keep up with the early theoretical work and reach the regime where the back-action of photons caused by the interaction between the radiation pressure and the mechanical oscillator influences the optomechanical dynamics significantly, producing a plenty of new phenomena, like optomechanically induced transparency and the great change of light propagation. In 2010, Kippenberg *et al.* first experimentally achieved this optomechanically induced transparency in cavity optomechanical system [Kippenberg (2010)]. They demonstrated that a pump optical beam tuned to a sideband transition of a micro-optomechanical system leads to destructive interference for the excitation of an intracavity probe field, inducing a tunable transparency window for the probe beam.

And then, less than half a year, in March 10, 2011, Teufel and colleagues at National Institute of Standards and Technology experimentally realized circuit cavity electromechanics in the strong-coupling regime, by incorporating a free-standing, flexible aluminium membrane into a lumped-element superconducting resonant cavity [Teufel (2011)]. They claimed that a parametric drive tone at the different frequency between the mechanical oscillator and the cavity resonance dramatically increases the overall coupling strength, which allows the complete entrance of the quantum-enabled, strong-coupling regime.

Afterwards, one month later, Painter *et al.* [Safavi-Naeini (2011)] at California Institute of Technology reported an optically tunable delay of 50 nanoseconds with near-unity optical transparency, and superluminal light with a 1.4 microsecond signal advance, in a nanoscale optomechanical crystal scheme. Via engineered photon-phonon interactions, they experimentally demonstrated that the optomechanical nonlinearity was used to control the velocity of light propagation at low temperatures. The physical origin of these results is due to the interaction between probe light and the intracavity photons, which gives rise to a large contribution to the phase dispersion. When the probe light passes through, the intracavity photons scattered and change the transmitting time of the probe light [Zhu and Li (2010b); Safavi-Naeini (2011)]. Furthermore, Wang's group [Fiore (2011)] recently implemented the light storage in a silica optomechanical resonator. They used the writing and readout laser pulses which tuned to the mechanical frequency below an optical cavity resonance to control the coupling between the mechanical displacement and the optical field.

Besides, many theoretical works that investigate the light propagation and quantum optical properties in the generalized optomechanical systems (especially solid-state optomechanical systems) are also taken by Zhu's group after his pioneering work in 2001 [Zhu and Li (2001)]. Since then, they also first proposed a scheme for measuring the coupling-rate of cavity and mechanical motion based on radiation pressure induced normal mode splitting and the vibrational frequency of mechanical resonator [Zhu and He (2010a)]. And then, they investigated the light propagation in a cavity optomechanical system with a Bose-Einstein condensate (BEC) [Chen and Zhu (2011)]. Their results show that the slow light can be realized in a BEC coupled to an optical cavity field under the radiation of two optical beams. Such relevant study also involves the tunable pulse delay and advancement device based on a cavity optomechanical system in microwave frequency range, the realization of nonlinear Kerr modulator, quantum optical transistor, quantum memory, and mass sensor in solid state optomechanical systems [Zhu and He (2010a); Zhu and Li (2010b, 2001); Chen and Zhu (2011); Jiang and Zhu (2012); Yuan and Zhu (2008); Chen and Zhu (2011); Jiang and Zhu (2010); Wang and Zhu (2010); Zhu and Li (2002); Jiang and Zhu (2006, 2008); Li and Zhu (2010a, 2009b, 2011c, 2009d, 2011f, 2010g, 2011h, 2010i, 2011j, 2010k, 2012l,m)]. All these theoretical effects need to be testified in the future and guide experiments to a convenient and explicit direction, i.e., the light propagation in solid state optomechanical system. Other recent theoretical works approach to cavity optomechanical system concerning about the light storage involving an optical waveguide coupled to an optomechanical crystal array [Painter (2011)], the laser phase noise on the cooling and heating of a generic cavity optomechanical system [Kanamoto (2010)], and the collective nonlinear optomechanical dynamics [Heinrich (2011)].

1.4   Hallmarks of optomechanical systems

The first hallmark of cavity optomechanical system in the ground state is the optical bistable behavior via optical pumping, which has been discussed theoretically [Kitano (1981)] and investigated in various systems since several decades ago [Kanamoto (2010); Brennecke (2008); Cecchi (1982); Joshi (2003)].

For certain values of pump-cavity detuning, it supports three steady-state solutions (Fig.1.2). A standard linear stability analysis shows that in the region with three solutions, two of them are dynamically stable, which represent the bistable behavior. We can calculate the energies of the stable states according to the Hamiltonian equation and identify the ground-state solutions [Kanamoto (2010); Brennecke (2008)], which are represented by the solid lines in Fig.1.2. For cavity optomechanical system, the ground state jumps from one branch to another at certain critical values as pump-cavity detuning is scanned [Dong (2011)]. These critical points correspond to first-order transitions in optomechanical systems. In Fig.1.2(a), increasing the pump rate results in a prominently tristable behavior, and

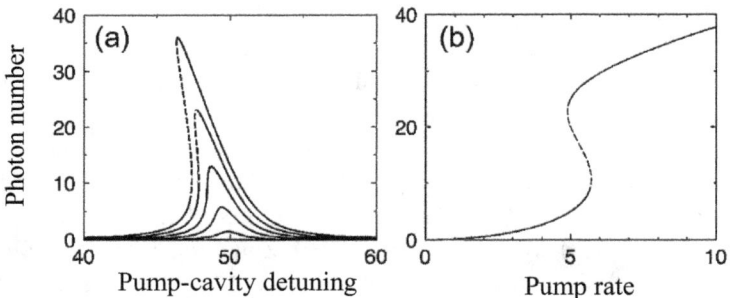

Fig. 1.2   The first hallmark of optomechanical system — bistable behavior. (a) Steady-state intracavity photon number as a function of (a) pump-cavity detuning for — by increasing maxima — pump rate (bistability threshold), (b) pump rate for different values of pump-cavity detuning [Kanamoto (2010)].

we observe that as the pump-cavity detuning decreases or increases even further, each pump-exciton detuning corresponds to each photon number, which leads to a stable state. It means that the input pump field directly affects the coupling between the cavity photons and the mechanical oscillator. Figure 1.2(b) shows the relationship between the photon number and the pump rate. The dashed line relates to the unstable behavior while the solid part corresponds to the stable state.

Furthermore, the optomechanical system can realize electromagnetically induced absorption (EIA), electromagnetically induced transparency (EIT) and parametric amplification (PA), under the manipulation of pump-probe technique. Such process is the second hallmark of optomechanical system. By applying the pump laser at a mechanical frequency away from cavity, i.e., on the red sideband, the EIT effect is presented, while on the blue sideband, we can observe EIA and PA phenomena simultaneously [Safavi-Naeini (2011)].

As the pump laser detuned from the cavity, the spectral selectivity of the optical cavity causes the sideband populations. One of the sidebands needs to be neglected, depending on whether the pump laser is on the red or blue sideband of the cavity. On the red sideband, the pump-cavity detuning $\Delta_{\mathrm{pu}} = \omega_c - \omega_{\mathrm{pu}} = \omega_m$ (where ω_c and ω_{pu} denote the frequency of cavity and pump laser, respectively), the total optomechanical system results in electromagnetically induced transparency, as occurring in atomic three-level systems, as shown in Ref.[Safavi-Naeini (2011)]. For the blue sideband $\Delta_{\mathrm{pu}} = -\omega_m$, there are two situations: (1) In the weak coupling of cavity-mechanical motion, the behavior of the optomechanical system is analogous to what has been observed in atomic gases, and the electromagnetically induced absorption occurs [Safavi-Naeini (2011); Lezama (1999)]. (2) As we tuned away from the weak cavity-mechanical coupling, the parametric amplification is always possible and occurs.

1.5 Generalized optomechanical systems

Lasers interacting with other atoms and molecules in electrical or optical pathways may stimulate the production of quantum interference. Electromagnetically induced transparency (EIT), observed in atomic materials, is a common quantum effect in Λ-style-systems, in which a pump laser induces a narrow spectral transparency window for a weak probe laser [Harris (1997); Lukin (2001)]. The dramatic change of the group velocity of probe field is caused by the rapid variation of the refractive index in this spectral window. Pump field regulates the magnitude of EIT window and enables a slow light, superluminal light and even a complete stop pulse of probe light in the atomic medium [Boyd (2002)].

Cavity optomechanics, the union of quantum optics and mechanics, is undergoing a time of academic pursuit. Similar with EIT, the transparency produced in cavity optomechanical system is the so called optomechanically induced transparency (OMIT) [Kippenberg (2010)]. The pump laser tuned to the lower motional sideband of the cavity resonance induces a dipole-like interaction between optical and mechanical degrees of freedom, enabling a tunable transparency window for the probe beam. This form of transparency is caused by radiation-pressure coupling of an optical mode and a mechanical mode. Dynamic control of OMIT via the pump laser in cavity optomechanical system produces some useful applications.

A typical optomechanical system generally consists of an optical cavity and a mechanical resonator, which exploits the interaction between photons and phonons via radiation pressure. For generalized optomechanical systems (GOS), the optical cavity is replaced by a two-level system, while the mechanical oscillator is replaced by the mechanical vibrations. After applying two optical fields, GOS exhibits the optomechanically induced transparency, as well as the electromagnetically induced absorption and parametric amplification. We conclude that either a two-level system or an optical cavity coupled to mechanical vibration that obeys the two hallmarks of optomechanics can form a generalized optomechanical system indeed.

Because of dressing with the mechanical vibration, the two-level system becomes three-level system and the whole system becomes transparent to the second probe laser, which is formally equivalent to EIT or OMIT situation. As shown in Fig.1.3(b), the two-level system is composed of the ground state $|g\rangle$ and the first excited state $|e\rangle$, which couples with the mechanical element. The mechanical element with the physical form of spring, has the similar dynamic behavior with harmonic oscillator. The two-level system can be optical materials such as optical fiber, photonic cavity, Bose-Einstein condensate, and the solid state materials like quantum dot, Cooper-pair box etc. where the lowest two energy levels play the main role. Because of dressing with the mechanical element, the two-level system exhibits Λ state (as shown in Fig.1.3(b)), which is similar with the EIT effect. Afterwards, a strong pump beam tuned to a sideband transition of a generalized optomechanical system leads to destructive interference for the excitation of an intra-cavity field, inducing a tunable transparency window for the probe beam (Fig.1.3(c)).

(a)

(b)

(1)   (2)   (3)

(c)

Probe-cavity detuning Δ_{pr} (MHz)

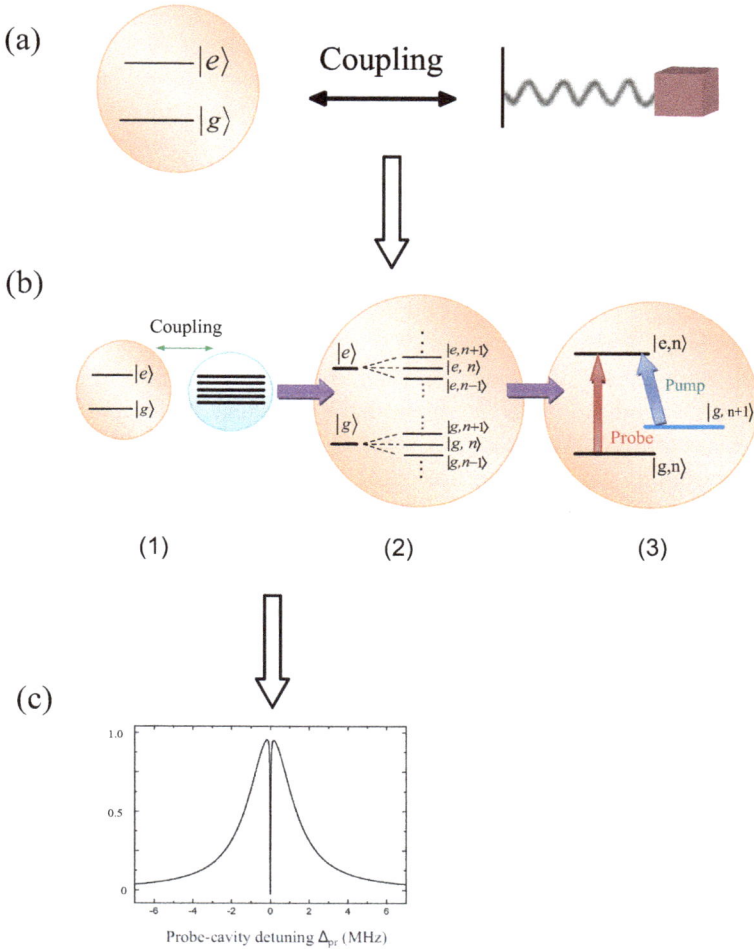

Fig. 1.3   (a) The two-level system coupled to mechanical vibration. (b) Since dressing with the mechanical element, the two-level system becomes Λ-type system. After radiating a strong pump laser and a weak probe laser, the dressed two-level system is similar with cavity optomechanical system. (1) The initial model of a two-level system and mechanical vibrations. $|g\rangle$ and $|e\rangle$ are the ground state and excited state of two-level system, respectively. (2) The initial energy levels of $|g\rangle$ and $|e\rangle$ split into several dressed states, when coupling the mechanical vibration. $|g, n\rangle = |g\rangle|n\rangle$ and $|e, n\rangle = |e\rangle|n\rangle$, where $|n\rangle$ denotes the number states of the mechanical vibrations. (3) The energy levels of $|g, n\rangle$, $|e, n\rangle$ and $|g, n + 1\rangle$ constitute the Λ-type structure, where the $|g, n + 1\rangle \rightarrow |e, n\rangle$ and $|g, n\rangle \rightarrow |e, n\rangle$ transitions can be induced by the pump beam and probe beam, respectively. (c) The absorption of the probe beam for the generalized optomechanical system, which signifies a transparency at $\Delta_{\text{pr}} = 0$ and matches well with typical optomechanical system.

The basic components of the generalized optomechanical system are as follows: (1) The mechanical element has the similar dynamic behavior with harmonic oscillator; (2) The two-level system is composed of the ground state $|g\rangle$ and the first excited $|e\rangle$. With these two elements, the investigated system can be treated as a generalized optomechanical system and can produce a host of long-anticipated phenomena similar with the typical cavity optomechanical systems.

Given above general description, it is no wonder that the trend of searching the transition from typical optomechanical systems to a variety of condense matter optomechanical systems, including superconducting microwave circuits [Jiang and Zhu (2010)], ultracold atoms [Kanamoto (2010)], Bose-Einstein condensate, photonic crystal cavity and other nanomaterials [Brennecke (2008)], which obeys the basic hallmarks of optomechanical system.

For example, Zhu *et al.* studied the parametric coupling in a optomechanical system consisting of a nanomechanical resonator and a superconducting microwave cavity [Jiang and Zhu (2010)]. Kanamoto and Meystre discussed theoretically the possibility of realizing optomechanical systems by using a degenerate gas of spinless fermions [Kanamoto (2010)]. They found that a mechanical fermionic mode analogous to the conventional moving mirror of optomechanics was provided by the collective density fluctuations associated with particle-hole excitations, and this mode can be quantized as a bosonic excitation. Furthermore, Gavartin *et al.* experimentally achieved the determination of the vacuum optomechanical coupling rate by using frequency modulation in optomechanical system, which consists of a suspended membrane containing a two dimensional photonic crystal defect cavity [Gavartin (2011)]. They declared that the strong coupling to the localized mechanical modes can also be changed to quantum dots in future studies. For the optomechanical system with Bose-Einstein condensate, Paternostro and colleagues recently investigated the dynamics and the possibility of an indirect diagnostic in an optomechanical system comprising of a cavity with a vibrating end mirror coupled to a Bose-Einstein condensate [Paternostro (2010)]. These extended studies pave the way towards the use of the mutual interaction between optics and solid state physics for the sake of coherent quantum control at the mesoscopic scale.

On the basis of above named studies, in this book, we discuss and analyze the propagation behaviors and quantum optical properties of the generalized optomechanical systems, under the radiation of a strong pump beam and a weak probe beam. We propose some condense matter materials based GOSs, which combine quantum optics and condensed matter physics together and hold promise for new optical sensitive devices, including semiconductor quantum dot, superconducting microwave cavity coupled to nanomechanical resonator, cavity with Bose-Einstein condensate system, etc. By dealing with the dynamic behaviors and light propagation characteristics of these condensed matter based GOSs, we predict some all-optical quantum devices, e.g., the slow light and fast light modulator, the nonlinear Kerr switch, the quantum optical transistor, the mass sensing and other quantum storage devices.

In Chapter 2, we first summarize the solution of basic Hamiltonian equation of motion with respect to GOS and put forward a set of special methods, including Heisenberg equation of motion, quantum Langevin equation and density matrix equations. By using any of these methods, linear susceptibility and nonlinear susceptibility of GOS with pump-probe technique can be obtained. We give a detailed calculation about how to solve the Hamiltonian of different GOS using these three methods, respectively, which provide an efficient way for people to master the quantum optical properties of GOS in a short time.

In Chapter 3, we present the light propagation of the typical cavity optomechanical systems and propose a precise way to measure the vacuum Rabi splitting of optical cavity and the vibrational frequency of mechanical resonator, due to radiation-pressure-induced normal mode splitting.

Taking the typical optomechanics as the foundation, Chapter 4-6 study the light propagation in three different generalized optomechanical systems, by using pump-probe technique. Chapter 4 investigates the light propagation in optical cavity containing a Bose-Einstein condensate, where the Bose-Einstein condensate is served as the mechanical vibrations. The nonlinear Kerr switch, slow light and fast light modulator, spectroscopic measurement and quantum optical transistor are presented in this system. Chapter 5 studies the smallest GOS with a single quantum dot (QD), the role of the optical cavity is played by an excitonic resonance of the QD, while the role of the mechanical element is played by the lattice vibrations. Light propagation in a single quantum dot gives rise to slow light, superluminal light, large optical Kerr effect on the red sideband and light amplification on the blue sideband, which offers a potential for a single quantum dot based quantum optomechanical transistor. The model in Chapter 6 is a single quantum dot embedded in nanomechanical resonator, where the role of optical cavity coupled to mechanical element is played by the interaction between exciton and the vibration of nanomechanical resonator. We demonstrate that the coupled quantum dot — nanomechanical resonator can be treated as a generalized optomechanical system. We present the light propagation characteristics and applications in this coupled system, such as the coherent optical spectroscopy, the vibrational frequency measurement, the transition from ultrafast light to ultraslow light, the quantum storage and memory etc. Chapter 7 shows a surface plasmon assisted optical mass sensing based on a hybrid metal nanoparticle-semiconductor quantum dot embedded in a doubly clamped nanomechanical resonator, driven by two optical fields. In Chapter 8, we investigate the optomechanical system with a carbon nanotube resonator. Especially, the optical mass sensing based on this carbon nanotube is also proposed in this section. Chapter 9 studies a nanomechanical resonator coupled a superconducting microwave cavity. Such system has been realized in experiments for the strong coupling and slow light achievement. The last Chapter is about a hybrid optomechanical system based on a quantum dot and DNA molecules. The physical protocol of the discrimination between abnormal cells and normal cells in optical domain is proposed in this section.

Apparently, there are other structures which can act as generalized optomechanical systems both classically and quantum mechanically. The most notable features are the two component elements: the two-level structure and the mechanical motion, as well as the two hallmarks of bistable behavior and EIT, EIA, PA effect, under the manipulation of pump-probe lasers. In this case, people can predict new generalized optomechanical systems by searching the basic two elements of optomechanics and the dressed states that can produce OMIT effect. Finally, we hope that the research about generalized optomechanical systems in this book will provide a explicit direction for experiments and applications in the future.

Bibliography

T. J. Kippenberg and K. J. Vahala, Cavity Optomechanics: Back-Action at the Mesoscale. Science 321, 1172 (2008).

T. J. Kippenberg, H. Rokhsari, T. Carmon, A. Scherer, and K. J. Vahala, Analysis of radiation-pressure induced mechanical oscillation of an optical microcavity. Phys. Rev. Lett. 95, 033901 (2005).

S. Gigan, H. R. Böhm, M. Paternostro, F. Blaser, G. Langer, J. B. Hertzberg, K. C. Schwab, D. Bäuerle, M. Aspelmeyer and A. Zeilinger, Self-cooling of a micromirror by radiation pressure. Nature 444, 67 (2006).

A. Schliesser, P. Del'Haye, N. Nooshi, K. J. Vahala, and T. J. Kippenberg, Radiation pressure cooling of a micromechanical oscillator using dynamical backaction. Phys. Rev. Lett. 97, 243905 (2006).

J. M. Ward, Y. Wu, V. G. Minogin, and S. Nic Chormaic, Trapping of a microsphere pendulum resonator in an optical potential. Phys. Rev. A 79, 053839 (2009).

A. Xuereb, T. Freegarde, P. Horak, P. Domokos, Optomechanical cooling with generalized interferometers. Phys. Rev. Lett. 105, 013602 (2010).

M. Aspelmeyer, S. Grölacher, K. Hammerer, and N. Kiesel, Quantum optomechanics-throwing a glance. J. Opt. Soc. Am. B 27, 6 (2010).

F. Marquardt, Optomechanics. Physics 2, 40 (2009).

K. J. Vahala, Optical microcavities. Nature 424, 839 (2003).

R. W. Boyd, Nonlinear Optics (Academic Press, Amsterdam) (2008).

R. W. Boyd and D. J. Gauthier, Controlling the Velocity of Light Pulses. Science 326, 5956 (2009).

G. S. Solomon, M. Pelton, Y. Yamamoto, Modification of spontaneous emission of a single quantum dot. Phys. Status Solidi 178, 341 (2000).

O. Painter, R. K. Lee, A. Scherer, A. Yariv, J. D. O'Brien, P. D. Dapkus, and I. Kim, Two-dimensional photonic band-gap defect mode laser. Science 284, 1819 (1999).

D. K. Armani, T. J. Kippenberg, S. M. Spillane, K. J. Vahala, Ultra-high-Q toroid microcavity on a chip. Nature 421, 925 (2003).

B. Gayral, J. M. Gérard, A. Lemaître, C. Dupuis, L. Manin, and J. L. Pelouard, High-Q wet-etched GaAs microdisks containing InAs quantum boxes. Appl. Phys. Lett. 75, 1908 (1999).

K. Djordjev, S. J. Choi, P. D. Dapkus, Microdisk tunable resonant filters and switches. IEEE Phot. Technol. Lett. 14, 828 (2006).

P. Rabiei, W. H. Steier, C. Zhang and L. R. Dalton, Polymer micro-ring filters and modulators. J. Lightwave Technol. 20, 1968 (2006).

J. D. Thompson, B. M. Zwickl, A. M. Jayich, F. Marquardt, S. M. Girvin, and J. G. E. Harris, Strong dispersive coupling of a high-finesse cavity to a micromechanical membrane. Nature 452, 72 (2008).

K. Karrai, I. Favero, and C. Metzger, Doppler optomechanics of a photonic crystal. Phys. Rev. Lett. 100, 240801 (2008).

D. Kleckner and D. Bouwmeester, Sub-kelvin optical cooling of a micromechanical resonator. Nature 444, 75 (2006).

C. Genes, D. Vitali, P. Tombesi, S. Gigan, and M. Aspelmeyer, Ground-state cooling of a micromechanical oscillator: Comparing cold damping and cavity-assisted cooling schemes. Phys. Rev. A 77, 033804 (2008).

V. Braginsky and A. Manukin, Ponderomotive effects of electromagnetic radiation. Sov. Phys. JETP 25, 653 (1967).

V. B. Braginsky, A. B. Manukin, and M. Y. Tikhonov, Investigation of dissipative Ponderomotive effects of electromagnetic radiation. Sov. Phys. JETP 31, 829 (1970).

H. Okamoto, D. Ito, K. Onomitsu, H. Sanada, H. Gotoh, T. Sogawa, H. Yam-
aguchi, Vibration amplification, damping, and self-oscillations in micromechanical resonators induced by optomechanical coupling through carrier
excitation. Phys. Rev. Lett. 106, 036801 (2011).

S. Gröblacher, J. B. Hertzberg, M. R. Vanner, G. D. Cole, S. Gigan, K. C. Schwab,
M. Aspelmeyer, Demonstration of an ultracold micro-optomechanical oscillator in a cryogenic cavity. Nature Phys. 5, 485 (2009).

A. Schliesser, O. Arcizet, R. Rivière, G. Anetsberger, T. J. Kippenberg, Resolved-
sideband cooling and position measurement of a micromechanical oscillator
close to the Heisenberg uncertainty limit. Nature Phys. 5, 509 (2009).

T. Carmon, H. Rokhsari, L. Yang, T. J. Kippenberg, and K. J. Vahala, Temporal
behavior of radiation-pressure-induced vibrations of an optical microcavity
phonon mode. Phys. Rev. Lett. 94, 223902 (2005).

C. Metzger, M. Ludwig, C. Neuenhahn, A. Ortlieb, I. Favero, K. Karrai, F.
Marquardt, Self-induced oscillations in an optomechanical system driven
by bolometric backaction. Phys. Rev. Lett. 101, 133903 (2008).

F. Marquardt, J. G. E. Harris, and S. M. Girvin, Dynamical multistability induced
by radiation pressure in high-finesse micromechanical optical cavities. Phys.
Rev. Lett. 96, 103901 (2006).

A. Schliesser, Cavity optomechanics and optical frequency comb generation
with silica whispering-gallery-mode microresonators, Thesis, Ludwig-
Maximilians-Universität München (2009); http://edoc.ub.uni-muenchen.
de/10940.

G. S. Agarwal and S. Huang, Electromagnetically induced transparency in mechanical effects of light. Phys. Rev. A 81, 041803 (2010).

W. He, J. J. Li and K. D. Zhu, Coupling-rate determination based on radiation-
pressure-inducd normal mode splitting in cavity optomechanical systems.
Opt. Lett. 35, 339 (2010a).

W. He, J. J. Li, and K. D. Zhu, All-optical control of light group velocity with a
cavity optomechanical system. ArXiv: 1011.4993v1 (2010b).

J. M. Dobrindt and T. J. Kippenberg, Theoretical analysis of mechanical displacement measurement using a multiple cavity mode transducer. Phys.
Rev. Lett. 104, 033901 (2010).

K. Stannigel, P. Rabl, A. S. Sørensen, P. Zoller, and M. D. Lukin, Optomechanical
transducers for long-distance quantum communication. Phys. Rev. Lett.
105, 220501 (2010).

D. E. Chang, A. H. Safavi-Naeini, M. Hafezi and O. Painter, Slowing and stop-
ping light using an optomechanical crystal array. New J. Phys. 13, 023003
(2011).

Y. Chang, T. Shi, Y. Liu, C. P. Sun, and F. Nori, Multistability of electromagnetically induced transparency in atom-assisted optomechanical cavities. Phys.
Rev. A 83, 063826 (2011).

A. B. Bhattacherjee, Cavity quantum optomechanics of ultracold atoms in an
optical lattice: Normal-mode splitting. Phys. Rev. A 80, 043607 (2009).

S. Weis, R. Rivière, S. Deléglise, E. Gavartin, Ol. Arcizet, A. Schliesser, and T. J. Kippenberg, Optomechanically induced transparency. Science 330, 1520 (2010).

J. D. Teufel, D. Li, M. S. Allman, K. Cicak, A. J. Sirois, J. D. Whittaker, and R. W. Simmonds, Circuit cavity electromechanics in the strong-coupling regime. Nature 471, 204 (2011).

A. H. Safavi-Naeini, T. P. Mayer Alegre, J. Chan, M. Eichenfield, M. Winger, Q. Lin, J. T. Hill, D. E. Chang, and O. Painter, Electromagnetically induced transparency and slow light with optomechanics. Nature 472, 69 (2011).

V. Fiore, Y. Yang, M. C. Kuzyk, R. Barbour, L. Tian, H. Wang, Storing optical information as a mechanical excitation in a silica optomechanical resonator. Phys. Rev. Lett. 107, 133601 (2011).

K. D. Zhu and W. S. Li, Electromagnetically induced transparency due to exciton-phonon interaction in an organic quantum well. J. Phys. B: At. Mol. Opt. Phys. 34, L679 (2001).

B. Chen, C. Jiang, K. D. Zhu, Slow light in a cavity optomechanical system with a Bose-Einstein condensate. Phys. Rev. A 83, 055803 (2011).

C. Jiang, B. Chen, K. D. Zhu, Controllable nonlinear responses in a cavity electromechanical system. J. Opt. Soc. Am. B29, 220 (2012).

X. Z. Yuan, H. S. Goan, C. H. Lin, K. D. Zhu, Y. W. Jiang, Nanomechanical-resonator-assisted induced transparency in a Cooper-pair-box system. New. J. Phys. 10, 095016 (2008).

B. Chen, C. Jiang, K. D. Zhu, Tunable all-optical Kerr switch based on a cavity optomechanical system with a Bose-Einstein condensate. J. Opt. Soc. Am. B 28, 8 (2011).

C. Jiang, J. J. Li, W. He and K. D. Zhu, Parametric coupling between a nanomechanical resonator and a superconducting microwave cavity. Europhys. Lett. 91, 58002 (2010).

H. Wang, K. D. Zhu, Coherent optical spectroscopy of a hybrid nanocrystal complex embedded in a nanomechanical resonator. Opt. Express, 18, 16175 (2010).

K. D. Zhu and W. S. Li, Electromagnetically induced transparency mediated by phonons in strongly exciton-phonon systems. Appl. Phys. B 76, 1 (2002).

Y. W. Jiang, K. D. Zhu, Z. J. Wu, X. Z. Yuan, M. Yao, Electromagnetically induced transparency in quantum dot systems. J. Phys. B: At. Mol. Opt. Phys. 39, 2621 (2006).

Y. W. Jiang and K. D. Zhu, Local field effects on phonon induced transparency in quantum dots embedded in a semiconductor medium. Appl. Phys. B 90, 79 (2008).

J. J. Li, K. D. Zhu, A tunable optical Kerr switch based on a nanomechanical resonator coupled to a quantum dot. Nanotechnol. 21, 205501 (2010a).

J. J. Li, K. D. Zhu, An efficient optical knob from slow light to fast light in a coupled nanomechanical resonator-quantum dot system. Opt. Express 17, 19874-19881 (2009b).

J. J. Li, K. D. Zhu, A quantum optical transistor with a single quantum dot in a photonic crystal nanocavity. Nanotechnol. 22, 055202 (2011c).

J. J. Li, K. D. Zhu, A scheme for measuring vibrational frequency and coupling strength in a coupled nanomechanical resonator-quantum dot system. Appl. Phys. Lett. 94, 063116 (2009); 94, 249903 (2009d).

J. J. Li, K. D. Zhu, All-optical Kerr modulator based on a carbon nanotube resonator. Phys. Rev. B 83, 115445 (2011e).

J. J. Li, K. D. Zhu, Quantum memory for light with a quantum dot system coupled to a nanomechanical resonator. Quantum Inf. Comput. 11, 0456-0465 (2011f).

J. J. Li, K. D. Zhu, Mechanical vibration-induced coherent optical spectroscopy in a single quantum dot coupled to a nanomechanical resonator. J. Phys. B 43, 155504 (2010g).

J. J. Li, W. He, K. D. Zhu, A quantum optomechanical transistor based on a cavity-optomechanical system. Arxiv: 1105.5753v1 (2011h).

J. J. Li, K. D. Zhu, Coherent optical spectroscopy due to lattice vibrations in a single quantum dot. Eur. Phys. J. D 59, 305 (2010i).

J. J. Li, K. D. Zhu, Plasmon-assisted mass sensing in a hybrid nanocrystal coupled to a nanomechanical resonator. Phys. Rev. B 83, 245421 (2011j).

J. J. Li, K. D. Zhu, A tunable optical Kerr switch based on a nanomechanical resonator coupled to a quantum dot. Nanotechnol. 21, 205501 (2010k).

J. J. Li, K. D. Zhu, Tunable slow and fast light device based on a carbon nanotube resonator. Opt. Exp. 20, 5840 (2012l).

J. J. Li, K. D. Zhu, Coherent optical spectroscopy in a biological semiconductor quantum dot-DNA hybrid system. Nano Res. Lett. 7, 133 (2012m).

G. A. Phelps, P. Meystre, Laser phase noise effects on the dynamics of optomechanical resonators. Phys. Rev. A 83, 063838 (2011).

G. Heinrich, M. Ludwig, J. Qian, B. Kubala, Collective dynamics in optomechanical arrays. Phys. Rev. Lett. 107, 043603 (2011).

M. Kitano, T. Yabuzaki, and T. Ogawa, Optical tristability. Phys. Rev. Lett. 46, 926 (1981); Self-sustained spin precession in an optical tristable system. Phys. Rev. A 24, 3156 (1981).

R. Kanamoto and P. Meystre, Optomechanics of a quantum-degenerate Fermi Gas. Phys. Rev. Lett. 104, 063601 (2010).

F. Brennecke, S. Ritter, T. Donner, T. Esslinger, Cavity optomechanics with a Bose-Einstein condensate. Science 322, 235 (2008).

S. Cecchi, G. Giusfredi, E. Petriella, and P. Salieri, Observation of Optical Tristability in Sodium Vapors. Phys. Rev. Lett. 49, 1928 (1982).

A. Joshi and M. Xiao, Optical multistability in three-level atoms inside an optical ring cavity. Phys. Rev. Lett. 91, 143904 (2003).

Y. Dong, J. Ye and H. Pu, Multistability in an optomechanical system with a two-component Bose-Einstein condensate. Phys. Rev. A 83, 031608 (2011).

A. Lezama, S. Barreiro, and A. M. Akulshin, Electromagnetically induced absorption. Phys. Rev. A 59, 4732 (1999).

S. E. Harris, Electromagnetically induced transparency. Phys. Today, 50, 36 (1997).

M. D. Lukin, A. Imamoğlu, Controlling photons using electromagnetically induced transparency. Nature 413, 273 (2001).

R. W. Boyd and D. J. Gauthier, Progress in Optics, edited by E. Wolf (Elsevier, Amsterdam), Vol. 43, pp. 497-530 (2002).

E. Gavartin, R. Braive, I. Sagnes, O. Arcizet, A. Beveratos, T. J. Kippenberg, I. Robert-Philip, Optomechanical coupling in a two-dimensional photonic crystal defect cavity. Phys. Rev. Lett. 106, 203902 (2011).

M. Paternostro, G. D. Chiara, and G. M. Palma, Cold-atom-induced control of an optomechanical device. Phys. Rev. Lett. 104, 243602 (2010).

Chapter 2

Theoretical Treatments in Generalized Optomechanical Systems

For light propagation in specific media, we are interested in the evolution of the variables associated with the system, which requires us to obtain the equations of motion for the system and enables the potential applications for more complicated quantum systems. There are several different approaches to deal with this problem. In this chapter, the main purpose is to offer readers a basic concept of how to solve the equations of motion in generalized optomechanical systems (GOS), and eventually predict the light propagation properties without figuring out the detailed results.

Here, we present three theoretical methods addressing to the equations of motion in generalized optomechanical system, which constitutes of a two-level system coupled to mechanical elements, in the presence of two optical fields. The damping with external reservoir is treated in different ways. One of the simplest semiclassical treatments is the use of Heisenberg equation of motion to deal with the equations of motion in generalized optomechanical systems. However, considering the damping with reservoir, the density matrix approach and the noise operator method are provided in the text. In density matrix treatment, the reservoir variables are eliminated by using the reduced density operator for the system in the Schrödinger picture, while in the noise operator method, the damping of the GOS is considered in the Heisenberg picture. We will demonstrate that all these three approaches are appropriate for the equations of motion to treat the light propagation properties of GOS.

In GOS, such as the quantum dot and the electron spin, a two-level description is valid if the two levels involved are resonant or nearly resonant with the driving fields, while the rest levels are highly detuned, which means the two levels possessing the lowest energy dominate the quantum process. The mechanical vibrations (e.g., mechanical resonator, Bose-Einstein

(a) Cavity optomechanical system

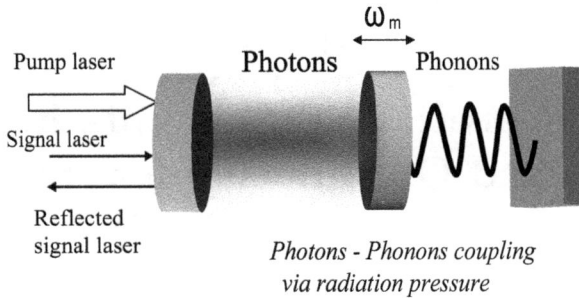

(b) Two-level system coupled to phonon modes

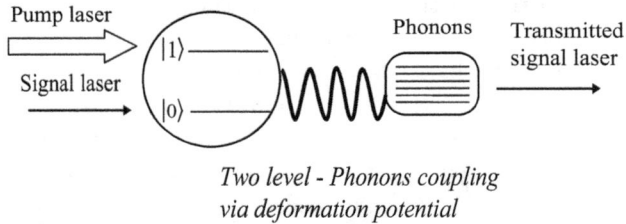

Fig. 2.1 Comparison of the conventional cavity optomechanics and GOS. (a) Typical cavity optomechanical system. The left mirror is fixed, while the right movable mirror couples to the optical cavity field via radiation pressure. (b) Generalized optomechanical system. The two-level system interacts with phonon modes via deformation potential in the presence of a strong pump laser and a weak signal laser. In contrast to the optomechanical system presented in (a), the light detection is the transmitted signal laser other than the reflected signal laser.

condensate, carbon nanotube, etc.) can be visualized as oscillators and be treated as phonon modes. Under certain realistic approximations, it is possible to reduce the GOS problem to an exact form which can be solved as two-level system coupled to phonon modes. The two-level system coupled to the phonon modes in GOS is analogous with the cavity photon-phonon coupling via radiation pressure in conventional cavity optomechanical system, as shown in Fig.2.1.

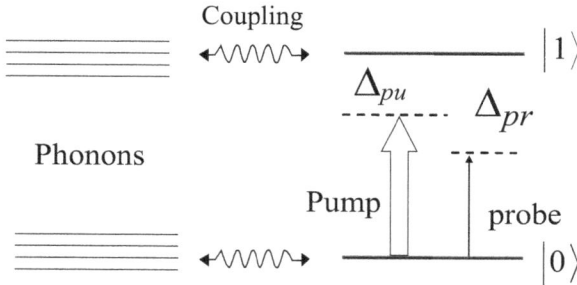

Fig. 2.2   The energy levels of two-level system while dressing with the mechanical modes (phonon modes), in the presence of a strong pump field and a weak probe field.

For simplicity, here we take a two-level system coupled to the mechanical modes as an example (Fig.2.2). The two-level system includes the ground state $|0\rangle$ and the first excited state $|1\rangle$, which can be characterized by the pseudospin $-1/2$ operators $S^{\pm}$ and S^z. The Hamiltonian of the two-level system can be described as

$$H_t = \hbar\omega_t S^z, \qquad (2.1)$$

where ω_t is the frequency of two-level system.

The mechanical element can be nano- or micro-mechanical resonators, which have the same eigenmode with quantum harmonic oscillator. We consider the specific case that the thickness of the resonator is smaller than its width, then the lowest-energy resonance corresponding to the fundamental flexural mode will constitute the resonator mode. We assume the resonator is characterized by high quality factors sufficiently [Imamoğlu (2004); Li (2009)]. The eigenmode of the mechanical element can be described by the bosonic annihilation and creation operators b and b^+, which has a quantum energy $\hbar\omega_n$, where ω_n is the resonant frequency. The Hamiltonian of the mechanical element is given by

$$H_n = \hbar\omega_n b^+ b. \qquad (2.2)$$

Generally speaking, the coupling to the mechanical resonator in generalized optomechanical systems seems complicated, due to the different coupling mechanisms between GOS elements. But all of them have the similar form of Hamiltonian, which can be described as follows. For a single quantum dot embedded in the mechanical resonator, the flexion of the resonator induces extensions and compressions, which will produce the longitudinal and the latitudinal strains. But only the longitudinal strain would modify the energy of the electronic states of quantum dot through deformation potential coupling. Then the Hamiltonian of the resonator mode coupled to the two-level quantum dot is described by

$$H_{t-n} = \hbar\omega_n \lambda S^z (b^+ + b), \qquad (2.3)$$

where λ represents the coupling between the mechanical resonator and quantum dot.

In the dipole approximation, when the field wavelength is larger than the two-level system size, the two-level system coupled to optical field can be mathematically equivalent to spin $-1/2$ particle interacting with a time-dependent electromagnetic field. The mechanical element is to modify the energy structure of two-level system via the deformation potential, which has nothing to do with the optical fields. Therefore, just as the spin $-1/2$ system undergoes the so-called Rabi oscillations between the spin-up and spin-down states under the action of optical fields, the two-level system also undergoes optical Rabi oscillations under the action of the driving optical fields, even considering the role of mechanical resonator. An understanding of this simple model enables us to consider more complicated quantum optical effects involving an ensemble of a two-level system interacting with optical fields. Figure 2.2 shows the energy levels of two-level system while dressing with the mechanical modes, driven by a strong pump field and a weak probe field. The Hamiltonian of a two-level system coupled with two optical fields can be given as

$$\begin{aligned}
H_{\text{int}} = &-\mu(S^+ E_{\text{pu}} e^{-i\omega_{\text{pu}}t} + S^- E_{\text{pu}}^* e^{i\omega_{\text{pu}}t}) \\
&-\mu(S^+ E_{\text{pr}} e^{-i\omega_{\text{pr}}t} + S^- E_{\text{pr}}^* e^{i\omega_{\text{pr}}t}),
\end{aligned} \qquad (2.4)$$

where E_{pu} and E_{pr} are slowly varying envelope of the pump field and probe field, respectively, μ is the electric dipole moment, ω_{pu} and ω_{pr} are the frequencies of pump field and probe field, respectively.

In the presence of a strong pump field and a weak probe field, the total Hamiltonian of GOS, including a two-level system, a mechanical element and their interacting with external optical fields can be written as:

$$\begin{aligned}
H = &\ H_t + H_n + H_{a-n} + H_{\text{int}} \\
= &\ \hbar\omega_t S^z + \hbar\omega_n b^+ b + \hbar\omega_n \lambda S^z (b^+ + b) \\
&-\mu(S^+ E_{\text{pu}} e^{-i\omega_{\text{pu}}t} + S^- E_{\text{pu}}^* e^{i\omega_{\text{pu}}t}) \\
&-\mu(S^+ E_{\text{pr}} e^{-i\omega_{\text{pr}}t} + S^- E_{\text{pr}}^* e^{i\omega_{\text{pr}}t}).
\end{aligned} \qquad (2.5)$$

In a frame rotating at the pump field frequency ω_{pu}, the Hamiltonian Eq.(2.5) can be written as:

$$
\begin{aligned}
H = {}& \hbar\Delta_{\text{pu}}S^z + \hbar\omega_n b^+ b + \hbar\omega_n \lambda S^z(b^+ + b) \\
& -\hbar(\Omega S^+ + \Omega^* S^-) - \mu(S^+ E_{\text{pr}} e^{-i\delta t} + S^- E_{\text{pr}}^* e^{i\delta t}),
\end{aligned}
\tag{2.6}
$$

where $\Delta_{\text{pu}} = \omega_t - \omega_{\text{pu}}$ is the frequency detuning between two-level system and pump field, $\delta = \omega_{\text{pr}} - \omega_{\text{pu}}$ is the pump-probe detuning, and $\Omega = \mu E_{\text{pu}}/\hbar$ is the Rabi frequency of the pump field.

In what follows, three theoretical treatments are presented to solve this Hamiltonian, i.e., Heisenberg equation of motion, density matrix approach, and quantum Heisenberg-Langevin approach. All these three methods will obtain the same results for light propagation properties of GOS.

2.1 Heisenberg equation of motion

According to the Heisenberg equation of motion

$$
i\hbar\frac{dO}{dt} = [O, H],
\tag{2.7}
$$

and the commutation relation

$$
[S^z, S^\pm] = \pm S^\pm, \quad [S^+, S^-] = 2S^z.
\tag{2.8}
$$

The temporal evolutions of the GOS, driven by two optical fields are given by

$$
\frac{dS^z}{dt} = i(\Omega S^+ - \Omega^* S^-) + \frac{i\mu E_{\text{pr}}e^{-i\delta t}}{\hbar}S^+ - \frac{i\mu E_{\text{pr}}^* e^{i\delta t}}{\hbar}S^-,
\tag{2.9}
$$

$$
\frac{dS^-}{dt} = -i\Delta_{\text{pu}}S^- - i\omega_n \lambda N S^- - 2i\Omega S^z - \frac{2i\mu E_{\text{pr}}e^{-i\delta t}}{\hbar}S^z,
\tag{2.10}
$$

$$
\frac{d^2 Q}{dt^2} + \omega_n^2 Q = -2\omega_n^2 \lambda S^z,
\tag{2.11}
$$

where $Q = b^+ + b$. Next we ignore the quantum properties of S^z, S^- and Q [Lam (1991); Zhu (2001)], and then the semiclassical equations read as follows:

$$\frac{dS^z}{dt} = -\Gamma_1(S^z + 1/2) + i\Omega S^+ - i\Omega^* S^-$$

$$+ \frac{i\mu E_{\mathrm{pr}}e^{-i\delta t}}{\hbar}S^+ - \frac{i\mu E_{\mathrm{pr}}^* e^{i\delta t}}{\hbar}S^-, \tag{2.12}$$

$$\frac{dS^-}{dt} = -(i\Delta_{\mathrm{pu}} + \Gamma_2)S^-$$

$$- i\omega_n \lambda N S^- - 2i\Omega S^z - \frac{2i\mu E_{\mathrm{pr}}e^{-i\delta t}}{\hbar}S^z, \tag{2.13}$$

$$\frac{d^2Q}{dt^2} + \gamma_n\frac{dQ}{dt} + \omega_n^2 Q = -2\omega_n^2\lambda S^z, \tag{2.14}$$

where we have introduced the damping terms phenomenologically [Boyd (2008)], so here Γ_1 is the relaxation rate of two-level system, Γ_2 is the dephasing rate of two-level system, and γ_n is the decay rate of the mechanical element due to the coupling to a reservoir of "background" modes and the other intrinsic processes [Imamoğlu (2004); Ekinci (2005); Wilson-Rae (2008)]. In order to solve Eqs.(2.12)-(2.14), we make the following ansatz [Boyd (2008)]

$$S^z(t) = S_0^z + S_+^z e^{-i\delta t} + S_-^z e^{i\delta t}, \tag{2.15}$$

$$S^-(t) = S_0 + S_+ e^{-i\delta t} + S_- e^{i\delta t}, \tag{2.16}$$

$$Q(t) = Q_0 + Q_+ e^{-i\delta t} + Q_- e^{i\delta t}. \tag{2.17}$$

By substituting Eqs.(2.15)-(2.17) in Eqs.(2.12)-(2.14) and working to the lowest order in E_{pr}, but to all orders in E_{pu}, we can obtain S_+ and S_-, which correspond to the linear and nonlinear optical susceptibilities, respectively:

$$\chi^{(1)}(\omega_{\mathrm{pr}}) = \frac{\mu S_+}{\epsilon_0 E_{\mathrm{pr}}} = \frac{\mu^2}{\epsilon_0\hbar\Gamma_2}\chi^{(1)}(\omega_{\mathrm{pr}}), \tag{2.18}$$

$$\chi^{(3)}(\omega_{\mathrm{pr}}) = \frac{\mu S_-}{3\varepsilon_0 E_{\mathrm{pu}}^2 E_{\mathrm{pr}}^*} = \frac{\mu^4}{3\varepsilon_0\hbar^3\Gamma_2^3}\chi^{(3)}(\omega_{\mathrm{pr}}), \tag{2.19}$$

where ϵ_0 is the dielectric constant of vacuum. These equations are valid for the generalized optomechanical systems.

2.2   Density matrix approach

In quantum optics, damping of a system is usually described by its interaction with a reservoir which has a large number of degrees of freedom.

However, we are only interested in the evolution of the variables associated with the system, which requires us to obtain the equations of motion for GOS after tracing over the reservoir variables.

In this subsection, we present a theory of damping based on the density operator in which the reservoir variables are eliminated by using the reduced density operator to deal with the equations of motion in GOS. For a two-level system, there are quantum dissipation associated with the vacuum state here, which are accompanied by fluctuations. The external fields may be visualized as a large number of harmonic oscillators. The coupling of a two-level system to a large number of oscillators leads to the decay, which reduces the energy distribution of the two-level system to a lower energy state. We now start with a general reservoir theory before considering the two-level system and the field are damping by a reservoir of harmonic oscillator (bosonic) modes.

We consider the decoherence and relaxation of two-level system and mechanical mode in combination with their interaction to external environment into the Hamiltonian [Gardiner (2000); Walls (1994); Carmichael (1999); Breuer (2002); Zhu (2008)]. Here we describe the environment as independent ensembles of harmonic oscillators, which have spectral densities. We also assume that the resonator interacts bilinearly with external environment via its position operators, and the two-level system interacts with the environment through S^x operator and S^z operator. The S^x coupling to the environment models the relaxation process of the two-level system, while the S^z coupling to the environment models the pure dephasing process of the two-level system [Gardiner (2000); Walls (1994); Carmichael (1999); Breuer (2002); Zhu (2008)]. On the other hand, because ω_t is much larger than ω_n, we use the rotating-wave approximation to the two-level and environment coupling term instead of the resonator-environment coupling term in the system-environment coupling Hamiltonian.

In accordance with standard procedure [Gardiner (2000); Walls (1994); Carmichael (1999); Breuer (2002); Zhu (2008)], we can obtain the Born-Markovian master equation of the reduced density matrix of the coupled system, $\rho(t)$, through tracing out the environmental degrees of freedom as

$$
\begin{aligned}
\frac{d\rho}{dt} = {}& -\frac{i}{\hbar}[H, \rho] + A\{[S^-, [S^+, \rho]] + h.c.\} \\
& + B\{[S^-, \{S^+, \rho\}] + h.c.\} + E[S^z, [S^z, \rho]] + D[Q, [Q, \rho]] \\
& + G[Q, [P, \rho]] + \frac{i}{\hbar}L[Q, \{P, \rho\}],
\end{aligned}
\tag{2.20}
$$

where $Q = -\lambda(b^+ + b)$ and P are the position and momentum operators of the resonator, respectively. The coefficients A, B, E, D, G and L correspond to the characteristics of the coupling, and to the structure and properties of the environment. Their explicit forms can be written as

$$A = -\frac{1}{2\hbar}\{\frac{\gamma_1}{2}(1 + 2N(\omega_t))\},$$

(2.21)

$$B = \frac{1}{2\hbar}\{\frac{\gamma_1}{2}\},$$

(2.22)

$$E = -\frac{1}{\hbar}\{\frac{\gamma_2}{2}(1 + 2N(0))\},$$

(2.23)

$$D = -\frac{1}{4\hbar}\gamma_3(1 + 2N(\omega_n)),$$

(2.24)

$$G = -\frac{\Delta_3}{2\hbar m_n \omega_n},$$

(2.25)

$$L = -\frac{\gamma_3}{4m_n \omega_n},$$

(2.26)

where

$$\gamma_1 = 2\pi J_x(\omega_t),$$

(2.27)

$$\gamma_2 = 2\pi J_z(0),$$

(2.28)

$$\gamma_3 = 2\pi J_c(\omega_n),$$

(2.29)

$$\Delta_3 = \Theta \int_0^\infty d\omega \frac{J_c(\omega)}{\omega - \omega_n}(1 + 2N(\omega)),$$

(2.30)

and $N(\omega) = 1/[\exp(\hbar\omega/k_B T) - 1]$ is the Bose-Einstein distribution of the thermal equilibrium environment. m_n is the effective mass of the mechanical element. J_x, J_z and J_c describe the spectral densities of the respective environment coupled to the two-level system through S^x and S^z, and to the resonator through Q, respectively. Θ denotes the principal value of the argument.

According to the master Eq.(2.20), we can obtain the equation of motion for the expectation value of any physical operator O of the coupled system by calculating $\langle\dot{O}(t)\rangle = Tr[O\dot{\rho}(t)]$. We thus have

$$\frac{d}{dt}\langle S^z\rangle = -\Gamma_1(\langle S^z\rangle + \frac{1}{2}) + i\Omega(\langle S^+\rangle - \langle S^-\rangle)$$
$$+ \frac{i\mu E_{pr}(e^{-i\delta t}\langle S^+\rangle - e^{i\delta t}\langle S^-\rangle)}{\hbar},$$

(2.31)

$$\frac{d}{dt}\langle S^-\rangle = -(i\Delta_{\text{pu}} + \Gamma_2)\langle S^-\rangle - i\omega_n\lambda\langle QS^-\rangle - 2i\Omega\langle S^z\rangle$$
$$-\frac{2i\mu E_{\text{pr}}e^{-i\delta t}}{\hbar}\langle S^z\rangle, \tag{2.32}$$

$$\frac{d^2}{dt^2}\langle Q\rangle + \gamma_n\frac{d}{dt}\langle Q\rangle + \omega_n^2\langle Q\rangle = -2\omega_n^2\lambda\langle S^z\rangle, \tag{2.33}$$

where Γ_1 and Γ_2 are the relaxation rate and dephasing rate of the two-level system, respectively. γ_n is the decay rate of the mechanical element. They are derived microscopically as

$$\Gamma_1 = \frac{2}{\hbar}\{\frac{\gamma_1}{2}(1 + 2N(\omega_t))\}, \tag{2.34}$$

$$\Gamma_2 = \frac{1}{\hbar}\{\frac{\gamma_1}{2}(1 + 2N(\omega_t))\} + \frac{4}{\hbar}\{\frac{\gamma_2}{2}(1 + 2N(0))\}, \tag{2.35}$$

$$\gamma_n = \frac{\gamma_3}{2m_n\omega_n}. \tag{2.36}$$

Note that if the pure dephasing coupling is neglected, i.e., $\gamma_2 = 0$, then $\Gamma_1 = 2\Gamma_2$. In order to solve these equations, we first take the semiclassical approach by factorizing the resonator and two-level system degrees of freedom, i.e., $\langle QS^-\rangle = \langle Q\rangle\langle S^-\rangle$, in which any entanglement between these systems should be ignored.

The rest of the solving processes are followed by Eqs.(15)-(19), which will obtain the same solutions for linear optical and nonlinear optical susceptibility of generalized optomechanical systems.

2.3 Quantum Heisenberg-Langevin equation

In the previous subsection, based on the density matrix approach, we developed the equation of motion for generalized optomechanical system, as it is evolved under the influence of reservoir systems. In this subsection, we consider the same problem of the GOS-reservoir variables. In addition to the damping terms, the noise operators in system's equations will produce fluctuations. These equations have the form of classical Langevin equations, which describe, the Brownian motion of a particle suspended in specific medium. The Heisenberg-Langevin approach discussed here is particularly suitable for the strong coupling regime of GOS.

By applying the Heisenberg equations of motion for operators S^z, S^- and Q and introducing the corresponding damping and noise terms, we derive the quantum Langevin equations as follows [Gardiner (2000); Walls (1994)]:

$$\frac{d}{dt}S^z = -\Gamma_1(S^z + \frac{1}{2}) + i\Omega S^+ - i\Omega^* S^- + \frac{i\mu E_{\text{pr}}e^{-i\delta t}}{\hbar}S^+$$
$$- \frac{i\mu E_{\text{pr}}^* e^{i\delta t}}{\hbar}S^-, \tag{2.37}$$

$$\frac{d}{dt}S^- = -(i\Delta_{\text{pu}} + \Gamma_2)S^- - i\omega_n\lambda QS^- - 2i\Omega S^z$$
$$- \frac{2i\mu E_{\text{pr}}e^{-i\delta t}}{\hbar}S^z + \hat{F}_e, \tag{2.38}$$

$$\frac{d^2}{dt^2}Q + \frac{1}{\tau_n}\frac{d}{dt}Q + \omega_n^2 Q = -2\omega_n^2\lambda S^z + \hat{\xi}, \tag{2.39}$$

where $\hat{F}_e$ is the δ-correlated Langevin noise operator, which has zero mean $\langle\hat{F}_e\rangle = 0$ and obeys the correlation function $\langle\hat{F}_e(t)\hat{F}_e^+(t')\rangle \sim \delta(t-t')$.

The motion of mechanical element is affected by thermal bath of Brownian and non-Morkovian process [Gardiner (2000); Giovannetti (2001)]. The quantum effects on the resonator are only observed in the limit of high quality factor, that obeys $Q = \omega_n/\gamma_n \gg 1$. The Brownian noise operator can be modeled as Markovian with the decay rate γ_n of the resonator mode. Therefore, the Brownian stochastic force has zero mean value $\langle\hat{\xi}\rangle = 0$ that can be characterized as [Giovannetti (2001)]

$$\langle\hat{\xi}^+(t)\hat{\xi}(t')\rangle = \frac{\gamma_n}{\omega_n}\int\frac{d\omega}{2\pi}\omega e^{-i\omega(t-t')}\left[1 + \coth\left(\frac{\hbar\omega}{2k_BT}\right)\right]. \tag{2.40}$$

Following standard methods from quantum optics, we derive the steady-state solution to Eqs.(2.37)-(2.39) by setting all the time derivatives to zero. They are given by

$$S_0^- = \frac{-2\Omega S_0^z}{(\Delta_{\text{pu}} + \omega_n\lambda Q_0) - i\Gamma_2}, \quad Q_0 = -2\lambda S_0^z, \tag{2.41}$$

where S_0^z determines the population inversion. To go beyond weak coupling, we can always rewrite each Heisenberg operator as the sum of its steady-state mean value and a small fluctuation with zero mean value as follows:

$$S^- = S_0^- + \delta S^-, S^z = S_0^z + \delta S^z, Q = Q_0 + \delta Q. \tag{2.42}$$

By inserting these equations into the Langevin Eqs.(2.37)-(2.39), one can safely neglect the nonlinear term $\delta Q\delta S^-$. Since the optical drives are weak, but classical coherent fields are not, we will identify all operators with their expectation values, and drop the quantum and thermal noise terms [Kippenberg (2010)]. Then the linearized Langevin equations can be written as:

$$\langle \delta \dot{S}^z \rangle = -\Gamma_1 \langle \delta S^z \rangle + i\Omega \langle \delta(S^-)^* \rangle - i\Omega^* \langle \delta S^- \rangle$$
$$+ \frac{i\mu E_{\text{pr}} e^{-i\delta t}}{\hbar} \langle \delta(S^-)^* \rangle - \frac{i\mu E_{\text{pr}}^* e^{i\delta t}}{\hbar} \langle \delta S^- \rangle, \qquad (2.43)$$

$$\langle \delta \dot{S}^- \rangle = -(i\Delta_{\text{pu}} + \Gamma_2) \langle \delta S^- \rangle - i\omega_n \lambda (\langle \delta S^- \rangle Q_0 + S_0^- \langle \delta Q \rangle)$$
$$- 2i\Omega \langle \delta S^z \rangle - \frac{2i\mu E_{\text{pr}} e^{-i\delta t}}{\hbar} \langle \delta S^z \rangle, \qquad (2.44)$$

$$\langle \delta \ddot{Q} \rangle + \frac{1}{\tau_n} \langle \delta \dot{Q} \rangle + \omega_n^2 \langle \delta Q \rangle = -2\omega_n^2 \lambda \langle \delta S^z \rangle. \qquad (2.45)$$

In order to solve Eqs.(2.43)-(2.45), we make the ansatz [Boyd (2008); Kippenberg (2010)]

$$\langle \delta S^- \rangle = S_+ e^{-i\delta t} + S_- e^{i\delta t}, \qquad (2.46)$$
$$\langle \delta S^z \rangle = S_+^z e^{-i\delta t} + S_-^z e^{i\delta t}, \qquad (2.47)$$
$$\langle \delta Q \rangle = Q_+ e^{-i\delta t} + Q_- e^{i\delta t}. \qquad (2.48)$$

Upon substituting the above ansatz into Eqs.(2.43)-(2.45), we can obtain the expressions of S_+ and S_-, which corresponding to linear and nonlinear optical susceptibilities, respectively.

In summary, if the Hamiltonian equation of motion of the system can be written in the form of Eq.(2.5), we can get the same results using any of these three methods, as discussed above. This means that one can predict the new materials or the new quantum optical properties of GOS without solving the detailed Hamiltonian equation, if handling these proposed theoretical methods proficiently.

After demonstrating the equations of motion of GOS, in the next several chapters, we shall introduce some specific generalized optomechanical systems, such as a single quantum dot, the cooper-pair box interacting with a nanomechanical resonator and carbon nanotubes, which all can be treated as two-level systems coupled to phonon modes. And these generalized optomechanical systems can realize phonon induced transparency (PIT), electromagnetically induced absorption (EIA) and parametric amplification (PA), which may serve as a quantum optomechanical transistor, an all-optical nonlinear modulator and so on.

Bibliography

I. Wilson-Rae, P. Zoller and A. Imamoğlu, Laser cooling of a nanomechanical resonator mode to its quantum ground state. Phys. Rev. Lett. 92, 075507 (2004).

J. J. Li and K. D. Zhu, A scheme for measuring vibrational frequency and coupling strength in a coupled nanomechanical resonator-quantum dot system. Appl. Phys. Lett. 94, 063116 (2009).

J. F. Lam, S. R. Forrest and G. L. Tangonan, Optical nonlinearities in crystalline organic multiple quantum wells. Phys. Rev. Lett. 66, 1614 (1991).

K. D. Zhu and W. S. Li, Electromagnetically induced transparency due to exciton-phonon interaction in an organic quantum well. J. Phys. B: At. Mol. Opt. Phys. 34, L679 (2001).

R. W. Boyd, *Nonlinear Optics* (Academic Press, Amsterdam) pp. 313 (2008).

K. L. Ekinci and M. L. Roukes, Nanoelectromechanical systems. Rev. Sci. Instrum. 76, 061101 (2005).

I. Wilson-Rae, Intrinsic dissipation in nanomechanical resonators due to phonon tunneling. Phys. Rev. B 77, 245418 (2008).

C. W. Gardiner, and P. Zoller, *Quantum Noise* (2nd edn) (Berlin: Springer) pp. 425-433 (2000).

D. F. Walls, and G. J. Milburn, *Quantum Optics* (Berlin: Springer) pp. 245-265 (1994).

H. Carmichael, *Statistical Methods in Quantum Optics I* (Berlin: Springer) pp. 261-268 (1999).

H.-P. Breuer, and F. Petruccione, *The Theory of Open Quantum Systems* (Oxford: Oxford University Press) pp. 441-497 (2002).

X. Z. Yuan, H. S. Goan, C. H. Lin, K. D. Zhu, Nanomechanical resonator assisted induced transparency in a Cooper-pair box system. New J. Phys. 10, 095016 (2008).

V. Giovannetti and D. Vitali, Phase-noise measurement in a cavity with a movable mirror undergoing quantum Brownian motion. Phys. Rev. A 63, 023812 (2001).

S. Weis, R. Rivière, S. Deléglise, E. Gavartin, Ol. Arcizet, A. Schliesser, *et al.* Optomechanically induced transparency. Science 330, 1520 (2010).

Chapter 3

Light Propagation in Cavity Optomechanical System

Optomechanics is a highly interdisciplinary subject that now receives enormous interest for its great potential in both fundamental and applied science [Favero (2012); Cole (2011); Nunnenkamp (2011)]. By coupling a high-finesse optical cavity to a low-frequency mechanical resonator, an optomechanical system makes many ultra-sensitive measurements possible and gives rise to dynamical back-action. Recently, strong coupling in such a system has been achieved, which paves the way towards full quantum optical control of light propagation and mechanical devices [Weis (2010); Teufel (2011); Safavi-Naeini (2011); Fiore (2011); Verhagen (2012)].

With the development of ultra-fast and ultra-small active components, all-optical devices with high-speed response time and good contrast ratios are required for future optical information processing and telecommunications. In this chapter, the light propagation properties of cavity optomechanical system are presented, such as the conversion from slow light to fast light, all-optical controlled quantum memory, measurement of vibrational frequency and vacuum Rabi splitting, and the optomechanical transistor. In view of the detailed theoretical calculations presented in Chapter 2, here we only introduce the light propagation properties, without any redundant equations. The physical process comes from the optomechanically induced transparency (OMIT), which has been demonstrated formally equivalent to electromagnetically induced transparency (EIT) in atomic three-level-systems and provides a new approach for controlling light pulses [Weis (2010); Kippenberg and Vahala (2008)]. Furthermore, EIT is an optical phenomenon that the absorption in an opaque medium can be eliminated via quantum interference, while retaining the large and desirable nonlinear optical properties [Harris (1990); Lukin (2001); Gauthier (2009)]. This phenomenon can be used to slow down light pulses, or even bring them to a complete stop.

We consider the typical experiment setup illustrated in Fig.1.1(a), where a Fabry-Perot cavity with effective length L is formed by one fixed partially transmitting and one movable perfectly reflecting mirror. A strong pump field with frequency of ω_{pu} and a weak probe field with frequency of ω_{pr} enter the cavity simultaneously through the fixed mirror, and then the circulating light gives rise to the radiation pressure force. In such a situation, the movable mirror is treated as a mechanical resonator with a frequency of ω_m, which is much smaller than the cavity free spectral range $c/2L$. Therefore we can adopt the single-cavity-mode description, since the scattering photons can be ignored as compared with that of the resonant mode ω_c. The Hamiltonian for this optomechanical system in the rotating frame is written as [He (2010)]

$$H = \hbar\Delta_{pu}a^+a + \frac{1}{2}\hbar\omega_m(p^2 + q^2) - \hbar Ga^+aq$$
$$+ i\hbar\sqrt{2\kappa}E_{pu}(a^+ - a) + i\hbar\sqrt{2\kappa}E_{pr}(a^+e^{-i\delta t} - ae^{i\delta t}), \qquad (3.1)$$

where $\Delta_{pu} = \omega_c - \omega_{pu}$ is the detuning of cavity and pump field, $\delta = \omega_{pr} - \omega_{pu}$ is the detuning of probe and pump field. The first term is the energy of the cavity mode, a and a^+ are the annihilation and creation operators for the cavity field, which obeys the commutation relation $[a, a^+] = 1$. The second term gives the energy of the mechanical mode, which is modeled as a harmonic oscillator and described by dimensionless position and momentum operators q and p with commutation relation $[q, p] = i$. The third term corresponds the coupling of the movable mirror to the cavity field via radiation pressure. The parameter $G = (\omega_c/L\sqrt{\hbar/m\omega_m})$ is the coupling rate between the cavity and the oscillator (m is the effective mass of the mechanical mode). The last two terms show the classical inputs with frequencies ω_{pu} and ω_{pr}, while E_{pu} and E_{pr} are respectively related to the laser power P by $|E_{pu}| = \sqrt{2P_{pu}\kappa/\hbar\omega_{pu}}$ and $|E_{pr}| = \sqrt{2P_{pr}\kappa/\hbar\omega_{pr}}$ (κ is the cavity amplitude decay rate).

According to the calculations described in Chapter 2, here for simplicity we use the Heisenberg equations of motion to deal with this cavity optomechanical system. The temporal evolution of the lowering operator a and the dimensionless position operator q are determined by the following semiclassical Heisenberg equations of motion, and are given by

$$\frac{da}{dt} = -(i\Delta_{pu} + \kappa)a + iGaq + \sqrt{2\kappa}(E_{pu} + E_{pr}e^{-i\delta t}), \qquad (3.2)$$

$$\frac{d^2q}{dt} + \gamma_m\frac{dq}{dt} + \omega_m^2q = \omega_m Ga^+a, \qquad (3.3)$$

where γ_m is the damping rate of mechanical mode. In order to solve these equations, we make the ansatz [Boyd (2008)] $a(t) = a_0 + a_+ e^{-i\delta t} + a_- e^{i\delta t}$, $q(t) = q_0 + q_+ e^{-i\delta t} + q_- e^{i\delta t}$. Upon substituting the approximation into Eqs.(3.2) and (3.3), and upon working to the lowest order in E_{pr} but to all orders in E_{pu}, and then we obtain in the steady state

$$a_+ = \sqrt{2\kappa} E_{pr} [\frac{-i\delta + (-i\Delta_{pu} + \kappa) + C}{(\kappa - i\delta)^2 + (\Delta_{pu} + iC)^2 - D}], \qquad (3.4)$$

$$a_- = |\sqrt{2\kappa} E_{pu}|^2 [\frac{iAB^* a_+^*}{F^2(i\delta + i\Delta + \kappa + C^*)}]. \qquad (3.5)$$

Here $A = G^2/\omega_m^2$, $B = \omega_m^2/(\omega_m^2 - i\gamma_m\delta - \delta^2)$, $C = iA\omega_m w_o + iAB\omega_m w_o$, $D = A^2 B^2 \omega_m^2 w_0^2$, $F = -(i\Delta + \kappa) + iA\omega_m w_0$ and $w_0 = |a_0|^2$. The parameter w_0 is determined by the equation:

$$w_0[\kappa^2 + (\Delta_{pu} - \frac{G^2}{\omega_m} w_0)^2] = |\sqrt{2\kappa} E_{pu}|^2. \qquad (3.6)$$

This form of cubic equation is characteristic of optical multistability [Gupta (2007); Brennecke (2008); Kanamoto (2010)]. Figure 3.1 shows the bistable behavior which is the hallmark of cavity optomechanical systems [Brennecke (2008)]. The dashed curve and solid curve represent the unstable state and stable state, respectively. The output field transmitted through the coupled system can be obtained using the standard input-output theory $a_{out}(t) = a_{in}(t) - \sqrt{2\kappa} a(t)$ [Zoller (2004)], where $a_{out}(t)$ is the output field operator. We have

$$\langle a_{out}(t) \rangle = a_{out0} + a_{out+} e^{-i\delta t} + a_{out-} e^{i\delta t} \qquad (3.7)$$

$$= \sqrt{2\kappa}(a_0 + a_+ e^{-i\delta t} + a_- e^{i\delta t})$$

$$= (E_{pu} - \sqrt{2\kappa} a_0)e^{-i\omega_{pu}t} + (E_{pr} - \sqrt{2\kappa} a_+)e^{-i(\delta + \omega_{pu})t}$$

$$- \sqrt{2\kappa} a_- e^{i(\delta - \omega_{pu})t}$$

$$= (E_{pu} - \sqrt{2\kappa} a_0)e^{-i\omega_{pu}t} + (E_{pr} - \sqrt{2\kappa} a_+)e^{-i\omega_{pr}t}$$

$$- \sqrt{2\kappa} a_- e^{-i(2\omega_{pu} - \omega_{pr})t}. \qquad (3.8)$$

We can see from Eq.(3.7) that the output field contains two input components (ω_{pu} and ω_{pr}) and one four-wave mixing (FWM) component at frequency $2\omega_{pu} - \omega_{pr}$. The transmission of the probe field, defined by the ratio of the output and input field amplitudes at the probe frequency, is

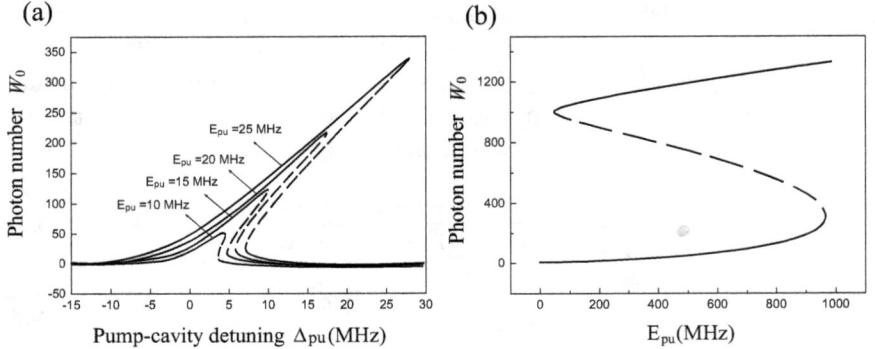

Fig. 3.1   The bistable behavior of cavity optomechanical system. (a) Steady-state intra-cavity photon number as a function of pump-cavity detuning, with different E_{pu}, where $E_{pu} = 25$MHz, $G = 0.9$MHz, $\kappa = 2\pi \times 215$kHz and $\omega_m = 10$MHz. (b) The mean intra-cavity photon number w_0 as a function of different values of E_{pu}, where $\Delta_{pu} = 30$MHz, $G = 0.9$MHz, $\kappa = 2\pi \times 215$kHz and $\omega_m = 10$MHz.

then given by

$$T = \frac{E_{pr} - \sqrt{2\kappa}a_+}{E_{pr}}$$
$$= 1 - 2\kappa \frac{-i\delta + (-i\Delta_{pu} + \kappa) + C}{(\kappa - i\delta)^2 + (\Delta_{pu} + iC)^2 - D}. \tag{3.9}$$

Figure 3.2(a) shows the probe transmission spectrum $|T|^2$ as a function of probe-cavity detuning $\Delta_{pr} = \omega_{pr} - \omega_c$. The spectrum curve splitting into two peaks can be interpreted in a dressed-state picture, as shown in Fig.3.2(b). The initial energy levels of cavity photon are dressed by the mechanical modes, via the radiation pressure. In Fig.3.2(b), the uncoupled energy levels on the left side split into dressed states $|N, n\rangle$ and the transitions between these states are related to one photon absorption or emission process. Therefore, the two peaks displaying sharp window in the probe spectrum correspond to the transitions between dressed states, and lead to a transparency at $\Delta_{pr} = 0$, which can be called optomechanically induced transparency (OMIT) [Weis (2010)]. Recently, Kippenberg and his research group [Weis (2010)] have experimentally demonstrated that OMIT operated in the resolved sideband regime is equivalent to EIT in a cavity optomechanical system.

There are several possible techniques for implementing light manipulation, such as electromagnetically induced transparency (EIT) [Tu (2009)], coherent population oscillation (CPO) [Xue (2010)], Raman and Brillouin amplification [Kalosha (2008)], among others [Elshaari (2010)]. EIT is the earliest technique to achieve the light storage in three-level system, which has been proved to be a powerful technique for eliminating the effect of a medium on a propagating beam of electromagnetic radiation [Harris (1997, 1990); Lukin (2003); Fleischhauer (2005)]. Since EIT retains the desirable and large nonlinear optical properties and the resonant response in material systems, it has been subjected to increasing investigation, especially in the realm of modern quantum optics and quantum information processing [Hu (2010); Hammerer (2010); Arikawa (2010)]. Here, different from EIT effect, the transparency occurred in cavity optomechanical system is due to optomechanics, which makes the excitation of an intracavity probe field become destructive interference, and eventually leads to a controlled transparency window of the input probe field [Huang (2010); Weis (2010)].

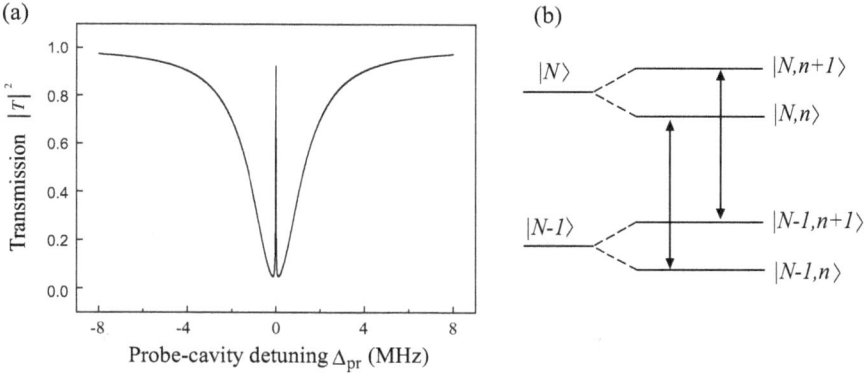

Fig. 3.2   (a) Probe transmission spectra as a function of probe-cavity detuning $\Delta_{pr} = \omega_{pr} - \omega_c$. The parameters used are $G = 0.9$MHz, $E_{pu} = 2$MHz, $\kappa = 2\pi \times 215$kHz, $\gamma_m = 2\pi \times 140$Hz, and $\Delta_{pu} = \omega_m = 10$MHz. (b) Energy levels and transitions of the cavity optomechanical system, where $|N\rangle$ and $|n\rangle$ represent the number states of cavity photon and mechanical modes, respectively.

3.1   Fast light and slow light

The tunable probe transmission window will modify the propagation dynamics of a probe pulse sent to this coupled system, due to the variation of the complex phase picked by its different frequency components. The probe pulse will experience a group delay τ_g, and this group delay τ_g is defined by

$$\tau_g = \frac{d\phi}{d\omega_{\rm pr}}\bigg|_{\omega_c} \tag{3.10}$$

where $\phi(\omega_{\rm pr}) = \arg(T(\omega_{\rm pr}))$ is the rapid phase dispersion. The magnitude and phase of the transmitted probe field could be determined experimentally by measuring the inphase and quadrature response of the system to the input modulation.

Before proceeding, we note that such an optomechanical system has an optical property which is analogous to the electrostriction or optical absorption in real material systems. As shown in Fig.1.1(a), when the pump field turns on, the circulating light illuminates on and gives rise to a radiation pressure force that deflects the mirror. In turn, the change of cavity's length alters the distribution of circulating intensity. This variation acts as an all-optical diffraction grating in the cavity field moving back and forth with the mechanical resonator oscillation frequency of ω_m. While the probe light travels in the cavity, the mutual interaction between input lights and the grating leads to the scattering of photons. If the probe light moves in the same direction as the diffraction grating, pump photons will be scattered into probe light, and hence a Stokes process occurs. On the contrary, if they move in different directions, probe light will be scattered into pump field, which results in an anti Stokes process. Such a behavior is very similar to the stimulated Brillouin scattering (SBS) in real material systems, in which an acoustic wave of frequency Ω is produced by the mutual interaction between light fields and material system. Through the process of electrostriction, the material system responses to the input fields by the fluctuations of dielectric constant which act as a moving diffraction grating with frequency Ω as shown in Fig.1.1(b) [Gauthier (2009); Boyd (2008)]. In view of Kramers-Kronig relations (K-K relation), the cavity optomechanical system will display a strong dispersive property while gain or loss resonance occurs, and therefore the control of light group velocity can be achieved.

Figure 3.3(a) and (b) plot both the transmission spectrum and the phase dispersion of the probe field. Parameters used in calculation are $E_{\rm pu} = 2\text{MHz}, \kappa = 2\pi \times 215\text{kHz}, \Delta_{\rm pu} = \omega_m = -10\text{MHz}$ and $\gamma_m = 2\pi \times 140Hz$ [Hammerer (2009)]. It is clear that the dispersion curve (Fig.3.3(b)) is very steep and positive around the center, which corresponds to a rapid variation in the group delay τ_g. At the same time the transmission spectrum splits into two peaks (normal mode splitting) at $\Delta_{\rm pr} = 0$ which ensures that the

(a)

(b)

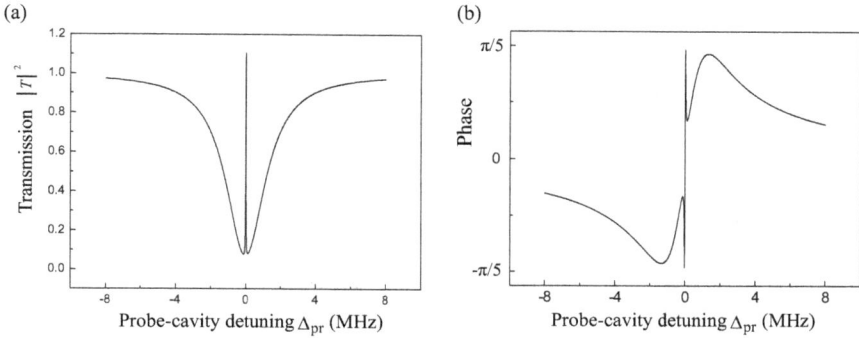

Fig. 3.3   Plot of the transmission spectrum (a) and the phase dispersion of probe field (b) as a function of probe-cavity detuning Δ_{pr}. Parameter values are $\Delta_{\mathrm{pu}} = -10\mathrm{MHz}$, $E_{\mathrm{pu}} = 2\mathrm{MHz}$, $\kappa = 2\pi \times 215\mathrm{kHz}$, $\gamma_m = 2\pi \times 140\mathrm{Hz}$ and $G = 1.2\mathrm{MHz}$.

probe light passes through with little energy loss. Hence, a large slow-light effect can occur around the center of the spectrum.

In Fig.3.4 we plot the group delay τ_g as a function of the pump power. The group delay of probe light is very sensitive to the pump power. The slope of the dispersion becomes steeper as P_{pu} decreases, leading to an increasingly lower group velocity. The physical origin of this result is due to the interaction between probe light and cavity field, as discussed above. The fluctuation of light intensity inside cavity acts as a moving diffraction grating composed of large amount of photons, and gives rise to a large contribution to the group delay. When the probe light passes through, the

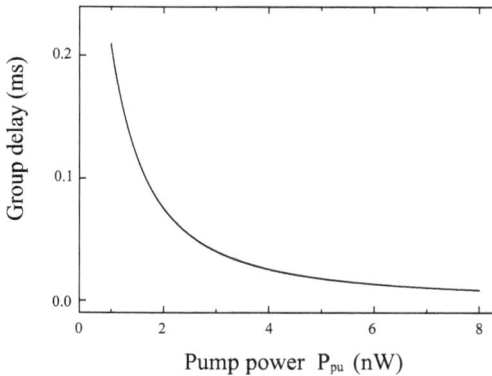

Fig. 3.4   The group delay τ_g of slow light versus P_{pu} with parameters $\Delta_{\mathrm{pu}} = -10\mathrm{MHz}$, $\kappa = 2\pi \times 215\mathrm{kHz}$, $\gamma_m = 2\pi \times 140\mathrm{Hz}$ and $G = 1.2\mathrm{MHz}$.

scattering of photons occurs which increases the transmitting time of the probe light.

From Eq.(3.7), we can get $a_{out+} = \sqrt{2\kappa}a_+$. For simplicity, we plot other two graphs in Fig.3.5, which shows the real part and imaginary part of a_{out+}. It should be noticed that these two curves are the same with Fig.3.3. The real part of a_{out+} exhibits absorptive behavior, and its imaginary part shows dispersive property, for the reason that the phase of light changes $\pi/2$ on the reflection. Therefore, in the following chapters, we can use $\text{Re}(a_{out+})$ and $\text{Im}(a_{out+})$ to substitute the Eq.(3.9).

Figure 3.6(a) and (b) give the transmission and phase dispersion part of the probe field respectively with the cavity-pump detuning fixed at 10MHz

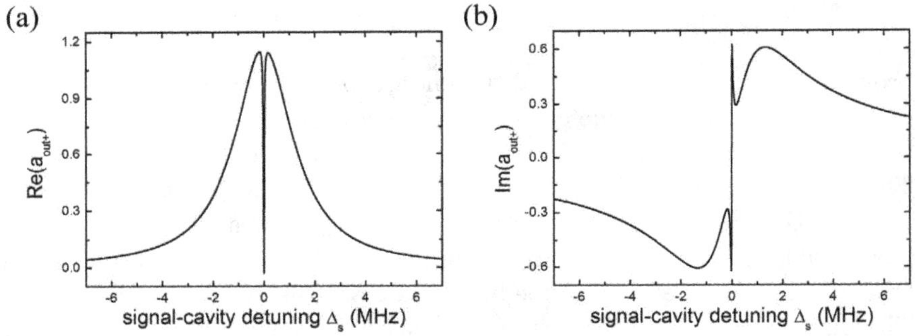

Fig. 3.5 Plot of the real part of a_{out+} (a) and the imaginary part of a_{out+} (b) as a function of probe-cavity detuning Δ_{pr}. Parameters values are the same with Fig.3.3.

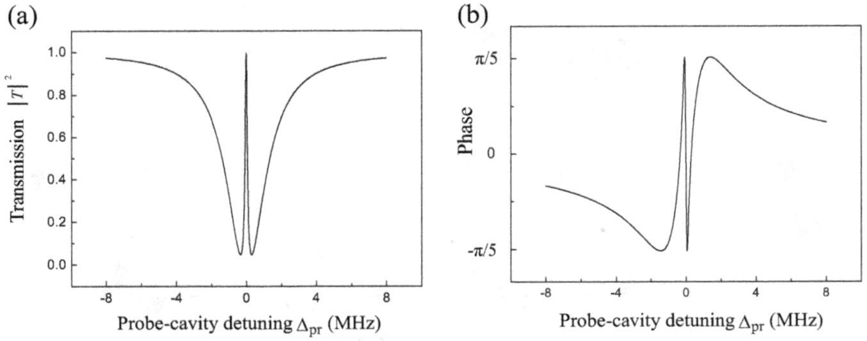

Fig. 3.6 (a) and (b) are the the transmission and phase dispersion part of the probe field respectively with $\Delta_{pu} = 10$MHz as a function of probe-cavity detuning. Other parameters are $E_{pu} = 5$MHz, $\kappa = 2\pi \times 215$kHz, $\gamma_m = 2\pi \times 140$Hz and $G = 1.2$MHz.

and with other parameters remained same as those in Fig.3.3. We find that the probe light also experiences a strong dispersive process, but the dispersion curve is negative which indicates a superluminal process occurs when the probe light travels in the cavity. And normal mode splitting in the transmission spectrum leads to transparency for the probe light. Figure 3.7 shows the variation of group delay with respect to the pump field power. The group delay of probe light is negative which means the pulse envelope travels backward in the system. This backward light also comes from the mutual interaction between probe light and cavity field. Compared with the slow light process, the difference is that the cavity-pump detuning is fixed at anti-Stokes peak, which determines the propagating direction of the moving grating, and further the optical scattering process of probe light inside cavity. Therefore it is possible to realize the switch between slow and superluminal light by simply choosing a proper cavity-pump detuning.

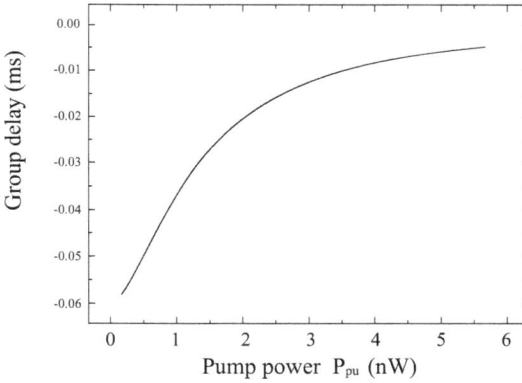

Fig. 3.7 The group velocity index $n_g = \frac{c}{v_g}$ of fast light versus E_{pu} with parameters $\Delta_{pu} = 10\text{MHz}$, $\kappa = 2\pi \times 215\text{kHz}$, $\gamma_m = 2\pi \times 140\text{Hz}$.

We also plot the real part and imaginary part of a_{out+} when $\Delta_{pu} = \omega_m = 10\text{MHz}$, which correspond to the absorption and dispersion of the probe field, respectively, as shown in Fig.3.8. It should be noticed that these two curves are the same with Fig.3.6.

3.2 All-optically controlled quantum memory

Storing and retrieving quantum light on demand, without corrupting the information it carries, have attracted wide attention in the field of quantum

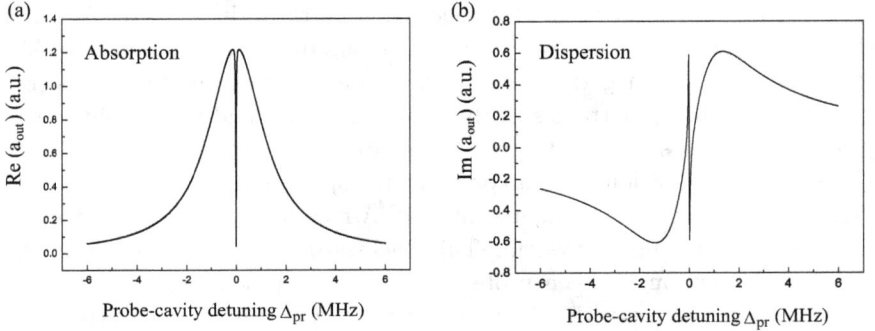

Fig. 3.8   Plot of the real part of a_{out+} (a) and the imaginary part of a_{out+} (b) as a function of probe-cavity detuning Δ_{pr}. Parameter values are $\Delta_{pu} = 10\text{MHz}$, $E_{pu} = 2\text{MHz}$, $\kappa = 2\pi \times 215\text{kHz}$, $\gamma_m = 2\pi \times 140\text{Hz}$ and $G_0 = 1.2\text{MHz}$.

optics and quantum information precessing [Riedmatten (2010); Hedges (2010); Karpa (2009)]. So far, successful demonstrations of quantum light storage have been used in atomic vapor [Dao (2010)], doped solid [Beil (2010)], and quantum cavity system [Elshaari (2010); Fiore and Yang (2011)]. Equally exciting is the possibility that these results will lead to useful applications in the fields of telecommunications and optical buffer [Lobino (2009)].

In this section, we predict that in cavity optomechanical system the quantum light memory can be achieved without any signal distortion. This is an all-optical controlled processing, in which the cavity photons interfere with the quantum probe pulse while dressing with the mechanical resonator mode (phonon mode). Theoretical analysis shows that light in the Fabry-Perot cavity can be dynamically and coherently transferred into long-lived vibrations of mechanical resonator, which provides a metastable state for photons. This dark-state polariton can be reaccelerated and converted back into a photon pulse by switching off and on the pump pulse on demand. Thanks to the long-lived mechanical resonator, the probe pulse can stay for a long time compared with other semiconductor systems. Our proposed quantum memory for light in an optomechanical system combines many of merits over other previous approaches, in which it allows the delay and release of pulse to be rapidly and all-optically controlled and achieves the long-term light memory due to the long lifetime of mechanical resonator. This scheme gives rise to the imagination of the ability of photons dressed by phonons, which can enable new possibilities for manipulating light through the manipulation of sound.

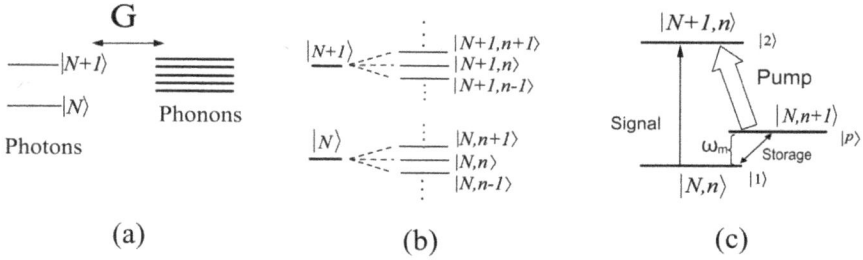

Fig. 3.9   The process of quantum memory for light (a) The initial model of the cavity photons and a mechanical resonator. G is the coupling between the optical cavity and mechanical resonator. The vibration modes of mechanical resonator are treated as phonons modes. (b) The split energy levels of the cavity photon when dressing the vibration mode of mechanical resonator. $|N\rangle$ and $|n\rangle$ denote the number states of the photon and phonon, respectively. (c) The process of quantum memory for light, where the $|N, n+1\rangle \to |N+1, n\rangle$ and $|N, n\rangle \to |N+1, n\rangle$ transitions can be induced by the pump beam and probe beam, respectively. $|N, n+1\rangle$ is the metastable state caused by mechanical vibration. The probe pulse is temporarily stored in the state between $|N, n+1\rangle$ and $|N, n\rangle$.

We consider the canonical situation in which a driven high-finesse optical cavity is coupled by momentum transfer of the cavity photons to a micro-mechanical resonator. The physical realization is shown in Fig.1.1(a). The cavity is driven by a strong pump field and a weak probe field with frequencies of ω_{pu} and ω_{pr}, respectively.

When the cavity photons couple to the mechanical resonator via the radiation pressure, the cavity photons will be dressed by an infinite number of possible phonon states as shown in Fig.3.9(b). Based on some experimental studies [Safavi-Naeini (2011); Weis (2010)], we can only select the three-level as a lambda structure as shown in Fig.3.9(c). All the metastable states other than $|N, n+1\rangle$ are not considered due to the transition energy of $|N, n\rangle \to |N+1, n\rangle$ is much higher than that in $|N, n\rangle \to |N, n+1\rangle$.

The mechanism of quantum memory by adiabatic passage is well known when the medium is made of lambda three-state atoms. The storage is implemented through the coherence of the atomic ground states. The idea of present work is to replace the lambda atoms by the dressed states picture as shown in Fig.3.9(c), and to take advantage of the long coherence induced by the resonator. We follow the treatment of [Fleischhauer (2002); Peng (2005)] and use a quasi-one-dimensional model, consisting of one propagating beam $\hat{\varepsilon}(z, t)$ passing through a cavity of length L. Here $\hat{\varepsilon}(z, t)$ is a weak probe field that couples the states $|N, n\rangle$ and $|N + 1, n\rangle$, and is related to the positive frequency part of the electric field by

$$\hat{E}_{\text{pr}}^{+}(z,t) = \sqrt{\frac{\hbar\omega_c}{2\varepsilon_0 V}}\hat{\varepsilon}(z,t)e^{i(\omega_c/c)(z-ct)}, \tag{3.11}$$

where ω_c is the frequency of the $|N,n\rangle \rightarrow |N+1,n\rangle$ transition (i.e., the frequency of the cavity photon), V is the quantization volume of the electromagnetic field, c is the light velocity in vacuum, ε_0 is the free space permittivity, and z is the direction along the cavity of length L. As compared with the three-level atomic systems, the $|N,n+1\rangle \rightarrow |N+1,n\rangle$ transition is driven by the strong pump laser (see Fig.3.9(c)). One can change the pump detuning simply by a tunable pump laser.

To perform a quantum analysis of the quantum probe light interacting with the cavity photons dressed by the resonator mode, it is useful to introduce locally averaged operators. $\aleph_z$ is the photon number in cavity along with z direction ($\aleph_z \gg 1$). Assuming the slowly-varying amplitude $\hat{\varepsilon}(z_j,t)$ does not change much, we introduce the locally-averaged, slowly-varying operators

$$\hat{\sigma}_{\mu\nu}(z_j,t) = \frac{1}{\aleph_{z_j}}\sum_{z_j \in N_z}\hat{\sigma}_{\mu\nu}^j(t)e^{i(\omega_{\mu\nu}/c)(z_j-ct)}, \tag{3.12}$$

where $\hat{\sigma}_{\mu\nu}^j(t) = |\mu^j(t)\rangle\langle\nu^j(t)|$ is the jth dressed photon state in cavity($\mu,\nu = 1,2,p$). For simplify, we use $|1\rangle$, $|p\rangle$, $|2\rangle$ to replace $|N,n\rangle$, $|N,n+1\rangle$, $|N+1,n\rangle$.

In the continuum limit, the effective interaction Hamiltonian for reduced three-level system can be written in terms of the locally-averaged operators as [Lukin (2003)]

$$\hat{H} = -\int\frac{\aleph\hbar}{L}[g\hat{\sigma}_{21}(z,t)\hat{\varepsilon}(z,t) + \Omega_{\text{pu}}\hat{\sigma}_{2p}(z,t) + \text{H.c.}]dz, \tag{3.13}$$

where $\aleph$ is the photon number density in Fabry-Perot cavity, $g = \mu\sqrt{\omega_c/2\varepsilon_0 V\hbar}$ is the cavity photon-quantum probe field coupling constant, $\Omega_{\text{pu}} = \langle n+1,N|\,\mu_{2p}\,E(t)\,|N+1,n\rangle/2\hbar$ describes the coupling to the pump field and the transition, $E(t)$ is the amplitude of the pump field, the "dipole moment" $\mu_{2\text{p}}$ corresponds to the transition between the state $|N+1,n\rangle$ and the state $|N,n+1\rangle$.

The evolution of the Heisenberg operator corresponding to the quantum probe field can be described in a slowly varying amplitude approximation by the propagation equation

$$(\frac{\partial}{\partial t} + c\frac{\partial}{\partial z})\hat{\varepsilon}(z,t) = ig\aleph\hat{\sigma}_{12}(z,t). \tag{3.14}$$

The dressed photon state evolution is governed by a set of Heisenberg-Langevin equations

$$\frac{\partial}{\partial t}\hat{\sigma}_{\mu\nu} = -\gamma_{\mu\nu}\hat{\sigma}_{\mu\nu} + \frac{i}{\hbar}[\hat{H},\hat{\sigma}_{\mu\nu}] + \hat{F}_{\mu\nu}, \tag{3.15}$$

where $\gamma_{\mu\nu}$ and $F_{\mu\nu}$ are the cavity decay rates and the δ-correlated Langevin noise operators, respectively.

Making the approximation that the quantum probe field intensity is much less than the pump field and assuming all the photon are initially in the state $|N,n\rangle$, we can solve Eq.(3.15) perturbatively to the first order in $g\hat{\varepsilon}/\Omega_{\text{pu}}$ to obtain a pair of two equations

$$\frac{\partial}{\partial t}\hat{\sigma}_{12} = -\gamma_{12}\hat{\sigma}_{12} + ig\hat{\varepsilon} + i\Omega_{\text{pu}}\hat{\sigma}_{1p} + \hat{F}_{12}, \tag{3.16}$$

$$\frac{\partial}{\partial t}\hat{\sigma}_{1p} = -\gamma_{1p}\hat{\sigma}_{1p} + +i\Omega_{\text{pu}}\hat{\sigma}_{12} + \hat{F}_{1p}, \tag{3.17}$$

where γ_{12} and γ_{1p} are the decay rates of $|N+1,n\rangle \to |N,n\rangle$ and $|N,n+1\rangle \to |N,n\rangle$, respectively. Assuming the pump laser power changed slowly and the decay rate γ_{1p} of the photon can be neglected, we can obtain [Fleischhauer (2002)]

$$\hat{\sigma}_{12}(z,t) \approx -\frac{i}{\Omega_{\text{pu}}}\frac{\partial}{\partial t}\hat{\sigma}_{1p}(z,t), \tag{3.18}$$

$$\hat{\sigma}_{1p}(z,t) \approx -g\frac{\hat{\varepsilon}(z,t)}{\Omega_{\text{pu}}}. \tag{3.19}$$

For optomechanical system, the propagation equation of quantum probe pulse in the perturbative and the adiabatic limit can be written as,

$$(\frac{\partial}{\partial t} + c\frac{\partial}{\partial z})\hat{\varepsilon}(z,t) = -\frac{g^2\bar{\aleph}}{\Omega_{\text{pu}}}\frac{\partial}{\partial t}\frac{\hat{\varepsilon}(z,t)}{\Omega_{\text{pu}}}, \tag{3.20}$$

where $\bar{\aleph}$ denotes the mean photon numbers in the cavity. The group velocity of probe field is given by

$$v_g = \frac{c}{1 + (\bar{\aleph}g^2/\Omega_{\text{pu}}^2)}. \tag{3.21}$$

In finite pump field and large $\bar{\aleph}$, the incident probe field can be slowed down significantly. One can obtain an analytical solution of Eq.(3.20) by introducing a new quantum field operator $\hat{\Psi}(z,t)$,

$$\hat{\Psi}(z,t) = \cos\theta(t)\hat{\varepsilon}(z,t) - \sin\theta(t)\sqrt{\bar{\aleph}}\hat{\sigma}_{1p}(z,t), \tag{3.22}$$

with

$$\cos\theta(t) = \frac{\Omega_{\text{pu}}(t)}{\sqrt{\Omega_{\text{pu}}^2(t) + g^2\bar{\aleph}}}, \qquad (3.23)$$

$$\sin\theta(t) = \frac{g\sqrt{\bar{\aleph}}}{\sqrt{\Omega_{\text{pu}}^2(t) + g^2\bar{\aleph}}}, \qquad (3.24)$$

$$\tan^2\theta(t) = \frac{g^2\bar{\aleph}}{\Omega_{\text{pu}}^2(t)}. \qquad (3.25)$$

Introducing the adiabaticity parameter $\varepsilon \equiv (g\sqrt{\bar{\aleph}}T)^{-1}$ with T being a characteristic time, one can expand the equations of motion in power of ε. In the lowest order, such as the adiabatic limit, we can obtain [Fleischhauer (2002)].

$$\hat{\varepsilon}(z,t) = \cos\theta(t)\hat{\Psi}(z,t), \qquad (3.26)$$

$$\sqrt{\bar{\aleph}}\hat{\sigma}_{1p} = -\sin\theta(t)\hat{\Psi}(z,t)e^{-i\Delta kz}. \qquad (3.27)$$

Furthermore, the new operator $\hat{\Psi}(z,t)$ obeys the following equation of motion

$$[\frac{\partial}{\partial t} + c\cos^2\theta(t)\frac{\partial}{\partial z}]\hat{\Psi}(z,t) = 0, \qquad (3.28)$$

which describes a shape-preserving propagation with velocity $v = v_g(t) = c\cos^2\theta(t)$. It is should be noted here that $\hat{\sigma}_{1p}(z,t)$ corresponds to the creation operator of the $|N,n\rangle \to |N,n+1\rangle$, which is different from atomic spin operator in three-level atomic systems [Lukin (2001)]. In the linear limit, this new operator satisfies the Bosonic commutation relation and we can refer this new Bosonic particle to phonon-polariton.

Equation (3.28) illustrates a shape- and quantum-state preserving propagation

$$\hat{\Psi}(z,t) = \hat{\Psi}(z,t)(z - c\int_0^t d\tau \cos^2\theta(\tau), 0). \qquad (3.29)$$

It is obvious that by adiabatically rotating θ from 0 to $\pi/2$ via a tunable pump laser, one can decelerate and stop the probe pulse on demand. When $\theta \to 0$, $\Omega_{\text{pu}}^2 \gg g^2\bar{\aleph}$, which corresponds to the strong external pump field, the polariton has purely photonic character $\hat{\Psi}(z,t) = \hat{\varepsilon}(z,t)$ and the probe propagation velocity equals to the vacuum speed of light. For $\theta \to \pi/2$, the polariton becomes phonon-like, $\hat{\Psi}(z,t) = -\sqrt{\bar{\aleph}}\hat{\sigma}_{1p}e^{i\Delta kz}$, and its propagation velocity approaches zero. In this process, the quantum probe field are mapped onto a phonon-polariton which are different from the atomic spins in three-level atomic systems. This phonon-polariton can then be reaccelerated to the vacuum speed of light in which the stored quantum states are transferred back to the photonic state. Figure 3.10 displays the coherent amplitude of a dark-state phonon-polariton, which results from an initial light pulse, as well as the corresponding field and matter components.

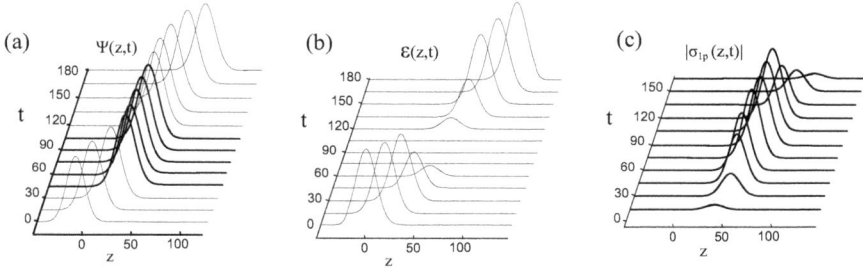

Fig. 3.10 Quantum light propagation of a dark-state polariton with envelope $\exp\{-(z/10)^2\}$, according to the expression $\cot\theta(t) = 100(1 - 0.5\tanh[0.1(t - 15)] + 0.5\tanh[0.1(t - 125)])$. (a) The coherent amplitude of the polariton $\hat{\Psi}(z,t) = \langle\hat{\Psi}(z,t)\rangle$ as functions of time t and length z. (b) The process of light storage. $\varepsilon(z,t) = \langle\hat{\varepsilon}(z,t)\rangle$. (c) The matter component part $|\sigma_{1p}| = |\langle\hat{\sigma}_{1p}\rangle|$, where $c = 1$.

Here, we show an example of the probe laser storage and recall on demand. First, under constant pump laser E_{pu}, the probe optical pulse E_{pr} completely enters the optomechanical system consisting of a Fabry-Perot cavity and a mechanical resonator (Fig.1.1(a)). Once the probe pulse inside, the amplitude of E_{pr} will turn to reduce gradually. However, when the pump laser is switched off rapidly, the probe laser information is stored in the coherence between the $|N, n\rangle$ and $|N, n + 1\rangle$ subbands. The maximum storage time is set by the mechanical decay rate, $\sim 1/\gamma_m$. Otherwise, turning on the pump laser results in a retrieved probe pulse. There is no additional distortion during the storage because the width of the probe pulse spectrum is very much less than the width of the OMIT window [Safavi-Naeini (2011)].

As we know, when storing light in atomic systems using EIT, characteristics of the field are recorded as a spin wave in the atomic ensemble. The storage time is determined by the coherence time of the hyperfine transitions. However, in optomechanical systems, the term *"storage"* implies the conversion of probe pulse into the phonon coherence σ_{1p}, whose lifetime is determined by the lifetime of mechanical resonator. Towards the realistic optomechanics systems, Gröblacher *et al.* [Hammerer (2009)] recently experimentally showed that in optomechanics system, the mechanical quality factor $Q = 6700$ and the decay rate $\gamma_m = 2\pi \times 140$Hz, which corresponds to the lifetime of resonator is 7ms. In this case, the light storage time can reach an ideal situation if we select a long vibration lifetime of mechanical resonator. Therefore, the optomechanical system we proposed here is a suitable choice for quantum light memory. Recently, Wang's group have experimentally realized the light storage in a silica optomechanical resonator [Fiore (2011)].

3.3   Measurement of vacuum Rabi splitting

Based on this optomechanical system, we can also simply measure the vacuum Rabi splitting. For illustration of the numerical results, we choose a realistic optomechanical system [Hammerer (2009)]. Distinct with Ref. [Hammerer (2009)], here we use two optical fields rather than the single probe optical detection. Figure 3.11(a) shows the probe transmission spectrum $|T|^2$ as a function of probe-cavity detuning $\Delta_{pr} = \omega_{pr} - \omega_c$ with different coupling rates. It shows that in the absence of coupling between cavity and mechanical oscillator, only a total absorption dip displayed in the probe transmission spectrum (solid curve). However, when the coupling turns on, the spectrum will split into two peaks, which can be interpreted in a dressed-state picture shown in Fig.3.2(b).

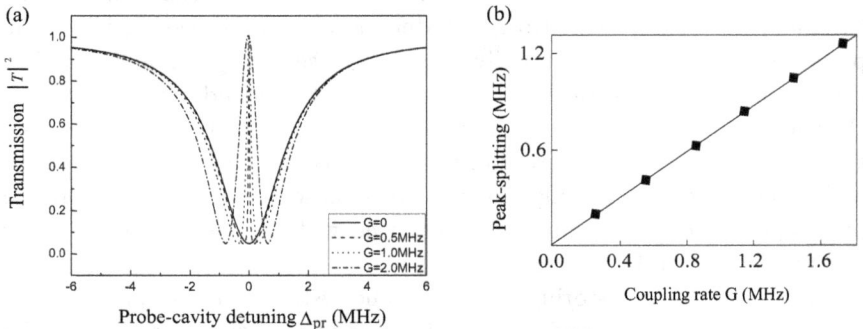

Fig. 3.11   (a) Probe transmission spectra as a function of probe-cavity detuning $\Delta_{pr} = \omega_{pr} - \omega_c$, with different coupling strengths G. The parameters used are $E_{pu} = 5$MHz, $\kappa = 2\pi \times 215$kHz, $\gamma_m = 2\pi \times 140$Hz, and $\Delta_{pu} = \omega_m = 10$MHz. (b) The relationship between peak-splitting and coupling strength.

Figure 3.11(b) shows the linear relationship between the distance between two splitting peaks and the coupling rates G. Figure 3.11(a) provides an effective and accurate method to measure the coupling rate between cavity and mechanical oscillator. Therefore, the coupling rate G, denoted as vacuum Rabi splitting, can be obtained by simply measuring the distance between two peaks in the probe transmission spectrum [He (2010)].

3.4 Measurement of resonator's frequency

To further investigate the property of this optomechanical system, we plot the probe transmission spectrum as a function of detuning Δ_{pr} with different ω_m, as shown in Fig.3.12. We find that the plot displays some new features. In the middle of the spectrum, there is a broad absorption dip, which corresponds to the resonance absorption of the cavity field. Also, for each value of ω_m, we can see another two sharp peaks at sidebands which exactly equals to $\pm\omega_m$. These two peaks represent the resonance absorption and amplification of the mechanical mode. Moreover, for the reason that the decay rate of mechanical oscillator is much smaller than the decay rate of cavity field, the spectral line of two sideband peaks is much

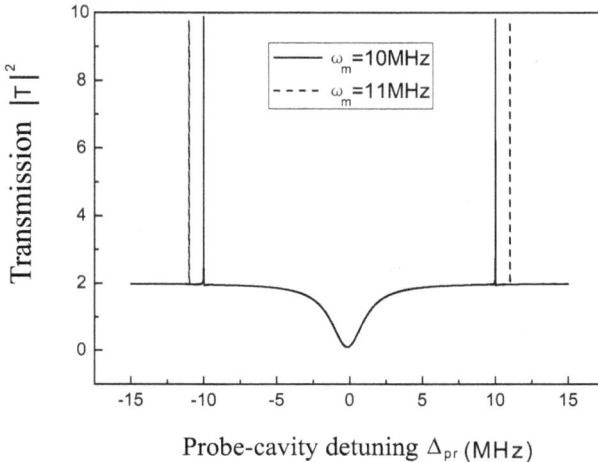

Fig. 3.12 The plot of the probe transmission spectrum with different vibrational frequencies of mechanical resonator. The parameters used are $E_{pu} = 1\text{MHz}$, $\kappa = 2\pi \times 215\text{kHz}$, $\gamma_m = 2\pi \times 140\text{Hz}$, $\Delta_{pu} = 0$, and $G = 0.9\text{MHz}$.

narrower than that of the peak in the center. Therefore, by scanning the probe frequency across the cavity frequency, it is convenient to obtain the frequency of mechanical mode in probe transmission spectrum, provided that the detuning $\Delta_{\mathrm{pu}} = 0$ [He (2010)].

3.5   An optomechanical transistor

In the pursuit of improved platforms for global information exchange and telecommunication, electrons can no longer serve as a signal carriers due to the heat effect, restricted transmission rate, energy loss, and the coherence between them [Orrit (2009)]. In this case, the continued increase in computing and communications highlights the need for new devices that reduce the effects of this electronic bottleneck by operating entirely in the optical domain [Miller (2010)]. Photons are the best choice to replace electrons as a signal carrier, which have robustness against decoherence and have a high transmission efficiency. One of the critical devices based on photons is the all-optical transistor — a device where a small optical 'gate' field is used to control the propagation of another optical 'signal' field via a specific processes [Tominaga (2001); Robinson (1984)] such as electromagnetically induced transparency (EIT) [Schmidt (2000); Raymond (2008); Bermel (2006)], optical Kerr effect [Moll (2006)], and cascaded second-order nonlinearity [Kim (1998)]. Recently, Hwang *et al.* have demonstrated that a single dye molecule can be operated as an optical transistor and coherently attenuate or amplify a tightly focused laser beam [Hwang (2009)].

Several materials and systems have been used to develop and demonstrate optical transistor including photonic crystal [Asakawa (2006); Yanik (2003)], atomic gas [Raymond (2008)], among others [Freitag (2010); Andrews (2010)]. However, these systems tend to be experimentally complex and most do not lend themselves to miniaturization. To this end, there has been a notable effort to create all-optical transistor using conveniently controlled manufacturing and miniaturization system. Actually, the general optomechanical system can serve as optomechanical transistor, in the presence of two optical fields. Here, in this section, we propose an all-optically controlled quantum optical transistor with toroidal nanocavity optomechanical system, as shown in Fig.3.13(a). Attracted by unique features of the optomechanical system, such as on-chip integration, large bandwidth of operation and long delay time, many theoretical and experimental researchers are focusing on its optical effects including the slow-light

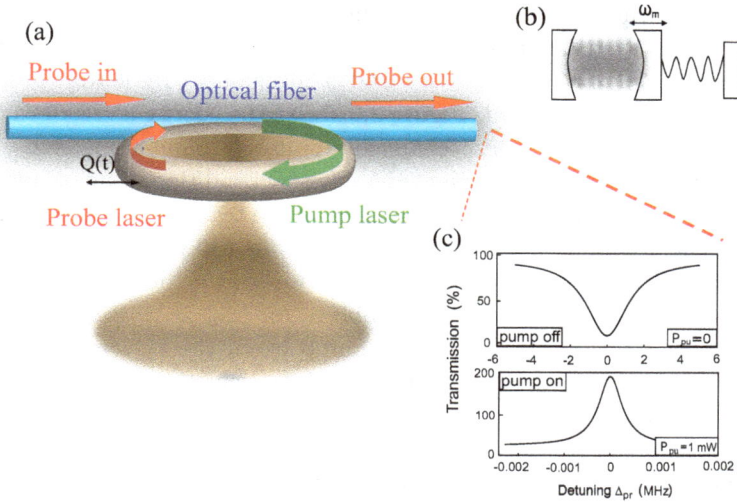

Fig. 3.13 Schematic diagram of a toroidal nanocavity-optomechanical system. (a) The input probe laser comes from the "in" port of optical fiber while the pump field exists in the nanocavity previously. Otherwise, the output probe laser can be detected at the "out" port of optical fiber. The nanocavity-resonator is coupled to the pump and probe fields using a optical fiber. (b) The corresponding typical optomechanical system. (c) Attenuation (pump off) and amplification (pump on) of the probe beam for the case $\Delta_{pu} = -10$MHz, $G = 0.9$MHz, $\omega_m = 10$MHz, $\kappa = 2\pi \times 215$kHz, $\gamma_m = 2\pi \times 140$kHz. $\mathcal{P}_{pu}$ is the pump laser power.

effect and fast-light effect [Safavi-Naeini (2011); Teufel (2011); Fiore (2011); Huang (2010)], normal splitting [He (2010)] and other effects [Singh (2010); Hofer (2010)]. We demonstrate that such an optomechanical system can indeed be served as a quantum optical transistor, where the cavity photons interfere with the input probe photons while the pump field controls the transmission of the probe laser. This quantum optomechanical transistor has the obvious merits that it can be switched on or off just by turning the pump beam on or off, which corresponds to amplification or attenuation of transmitted probe laser, respectively. The scheme proposed here will pave the way towards many important applications such as quantum communication and quantum repeaters.

In what follows, we consider the canonical situation in which a driven high-finesse optical cavity is coupled by momentum transfer of the cavity photons to a micro-mechanical resonator. The physical realization is shown in Fig.3.13(a), which a toroidal nanocavity is coupled to the mechanical radial breathing mode of the structure via radiation pressure. This toroidal nanocavity system can be modeled as a typical optomechanical system consisting of a Fabry-Perot cavity and a mechanical resonator as shown in Fig.3.13(b) [Weis (2010)]. Figure 3.9 shows the energy levels of cavity photons while dressing with mechanical vibrations (phonon modes).

In the following, we present the physical condition of quantum optomechanical transistor. Figure 3.13(c) shows the transmission spectrum of the probe field under the strong pump field. The top curve in Fig.3.13(c) displays the transmission spectrum when we shelve the pump beam in this transistor but only applying a probe beam ($\Delta_{pr} = \omega_{pr} - \omega_c$). This plot shows that in the absence of the pump beam the system attenuates the weak probe beam totally. This dip arises from the usual cavity absorption resonance. However, as turning on the pump beam and fixing the pump beam detuning $\Delta_{pu} = -\omega_m = -10\text{MHz}$, the dip becomes a peak immediately (see the bottom curve in Fig.3.13(c)).

As the pump power increases even further, we can observe more amplification of the probe beam as shown in Fig.3.14. Figure 3.14(a) shows the transmission spectra of the probe beam as a function of probe-cavity

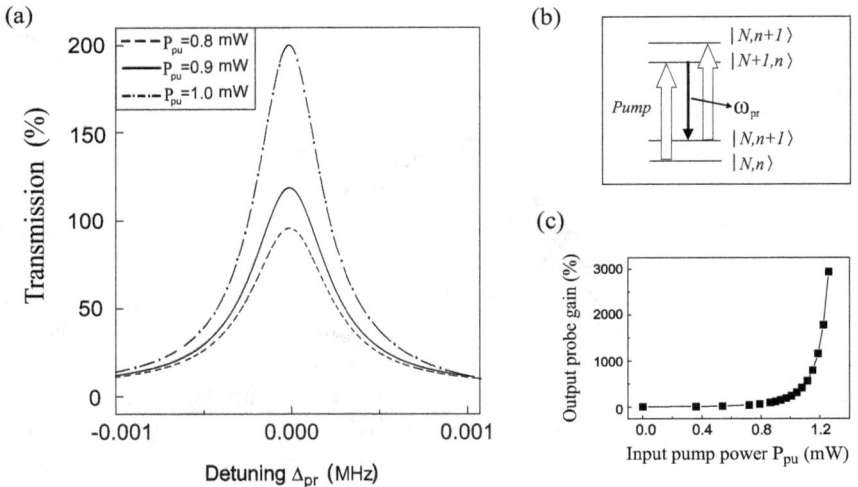

Fig. 3.14   The amplification of the probe beam with the same parameters as Fig.3.13(c). The transmission of the probe laser decreases as the power of pump beam increases. The energy levels and transitions are shown in the top inset. The bottom inset is the characteristic curve of quantum optomechanical transistor by plotting the transmission of the probe beam as a function of pump laser power.

detuning, which indicates that when the input pump power increases, the probe transmission increases rapidly. This result agrees well with the recent experiment in circuit micro-cavity electromechanics [Teufel (2011)]. This pump beam, just like a switch, dramatically controls the transmission spectrum of the probe beam. These plots in Fig.3.14(a) demonstrate that an optomechanical system can indeed act as a quantum optical transistor. This amplification behavior is caused by quantum interference between the dressed states while applying two optical fields. Figure 3.14(b) shows the origin of this three-photon resonance physical processing. Here the cavity photons makes a transition from the energy level $| N, n \rangle$ to the dressed level $| N + 1, n + 1 \rangle$ by the simultaneous absorption of two pump photons and emission of a probe photon at $\Delta_{pr} = 0$, as indicated by the region of amplification of the probe beam in Fig.3.14(a). In this case, optomechanics based optical transistor is an all-optical process, which is very different from atom and molecule optical transistor. Because of the all-optical physical situation, this optomechanical transistor can act as an optical router, which can provide enough energy for the transmitted probe laser while the light loses its energy during the transmission and makes the probe light amplified. For more specific description, we further investigate the transistor characteristic curve by plotting the amplification of the probe beam as a function of the input pump power as shown in Fig.3.14(c). In this case, the gain of the probe beam can be enhanced abruptly by increasing the input pump power, which is the typical behavior of the optical transistor. Therefore this cavity optomechanical system can be referred as a quantum optomechanical transistor.

Bibliography

I. Favero, Optomechanics: The stress of light cools vibration, Nature Phys. doi:10.1038/nphys2221 (2012).

G. D. Cole, M. Aspelmeyer, Cavity optomechanics: Mechanical memory sees the light: A nanomechanical beam coupled to an optical cavity can be operated as a non-volatile memory element, Nature Nanotechnol. 6, 690-691 (2011).

A. Nunnenkamp, K. Børkje, S. M. Girvin, Phys. Rev. Lett. 107, 063602 (2011).

S. Weis, R. Rivière, S. Deléglise, E. Gavartin, O. Arcizet, A. Schliesser, T. J. Kippenberg, Optomechanically induced transparency, Science 330, 1520 (2010).

J. D. Teufel, D. Li, M. S. Allman, K. Cicak, A. J. Sirois, J. D. Whittaker, R. W. Simmonds, Circuit cavity electromechanics in the strong-coupling regime, Nature 471, 204-208 (2011).

A. H. Safavi-Naeini, T. P. Mayer Alegre, J. Chan, M. Eichenfield, M. Winger, Q. Lin, J. T. Hill, D. E. Chang, and O. Painter, Electromagnetically induced transparency and slow light with optomechanics, Nature 472, 69-73 (2011).

V. Fiore, Y. Yang, M. C. Kuzyk, R. Barbour, L. Tian, and H. Wang, Storing optical information as a mechanical excitation in a silica optomechanical resonator, Phys. Rev. Lett. 107, 133601 (2011).

E. Verhagen, S. Deléglise, S. Weis, A. Schliesser, T. J. Kippenberg, Quantum-coherent coupling of a mechanical oscillator to an optical cavity mode, Nature 482, 63-67 (2012).

T. J. Kippenberg, and K. J. Vahala, Cavity optomechanics: back-action at the mesoscale, Science 321, 1172-1176 (2008).

S. E. Harris, J. E. Field, A. Imamoğlu, Nonlinear optical processes using electromagnetically induced transparency, Phys. Rev. Lett. 64, 1107-1110 (1990).

M. D. Lukin, A. Imamoğlu, Controlling photons using electromagnetically induced transparency, Nature 413, 273-276 (2001).

R. W. Boyd, and D. J. Gauthier, Controlling the velocity of light pulses, Sience 326, 1074-1077 (2009).

W. He, J. J. Li, K. D. Zhu, Coupling-rate determination based on radiation pressure-induced normal mode splitting in cavity optomechanical systems, Opt. Lett. 35, 339-341 (2010).

R. W. Boyd, Nonlinear Optics (Academic Press, Amsterdam) pp. 297-304 (2008).

S. Gupta, K. L. Moore, K. W. Murch, and D. M. Stamper-Kurn, Cavity nonlinear optics at low photon numbers from collective atomic motion, Phys. Rev. Lett. 99, 213601 (2007).

F. Brennecke, S. Ritter, T. Donner, and T. Esslinger, Cavity optomechanics with a Bose-Einstein condensate, Science 322, 235-238 (2008).

R. Kanamoto and P. Meystre, Optomechanics of a quantum-degenerate Fermi gas, Phys. Rev. Lett. 104, 063601 (2010).

C. W. Gardiner and P. Zoller, Quantum Noise (Springer) (2004).

Y. Tu, G. Zhang, Z. Zhai, and J. Xu, Angular multiplexing storage of light pulses and addressable optical buffer memory in $Pr^{3+} : Y_2SiO_5$ based on electromagnetically induced transparency, Phys. Rev. A 80, 033816 (2009).

W. Q. Xue, S. Sales, J. Capmany, and J. Mørk, Wideband 360° microwave photonic phase shifter based on slow light in semiconductor optical amplifiers, Opt. Express 18, 6156 (2010).

V. P. Kalosha, W. H. Li, F. Wang, L. Chen, and X. Y. Bao, Frequency-shifted light storage via stimulated Brillouin scattering in optical fibers, Opt. Lett. 33, 2848 (2008).

A. W. Elshaari, A. Aboketaf, and S. F. Preble, Controlled storage of light in silicon cavities, Opt. Express 18, 3014 (2010).

S. E. Harris, Electromagnetically induced transparency, Phys. Today 50 36 (1997).

M. D. Lukin, Colloquium: Trapping and manipulating photon states in atomic ensembles, Mod. Rev. Phys. 75, 457-472 (2003).

M. Fleischhauer, A. Imamoğlu, and J. P. Marangos, Electromagnetically induced transparency: Optics in coherent media, Rev. Mod. Phys. 77, 633 (2005).

X. M. Hu, H. Sun, and F. Wang, Scalable network of quadrangle entanglements via multiple phase-dependent electromagnetically induced transparency, Phys. Rev. A 82, 045807 (2010).

K. Hammerer, A. S. Sørensen, and E. S. Polzik, Quantum interface between light and atomic ensembles, Rev. Mod. Phys. 82, 1041-1093 (2010).

M. Arikawa, K. Honda, D. Akamatsu, S. Nagatsuka, K. Akiba, A. Furusawa, and M. Kozuma, Quantum memory of a squeezed vacuum for arbitrary frequency sidebands, Phys. Rev. A 81, 021605 (2010).

S. Huang, and G. S. Agarwal, Normal-mode splitting and antibunching in Stokes and anti-Stokes processes in cavity optomechanics: Radiation-pressure-induced four-wave-mixing cavity optomechanics, Phys. Rev. A 81, 033830 (2010).

S. Gröblacher, K. Hammerer, M. R. Vanner and M. Aspelmeyer, Observation of strong coupling between a micromechanical resonator and an optical cavity field, Nature 460, 724 (2009).

H. de Riedmatten, Quantum optics: Light storage at record bandwidths, Nat. Photon. 4, 206 (2010).

M. P. Hedges, J. J. Longdell, Y. M. Li, and J. Sellars, Efficient quantum memory for light, Nature 465, 1052 (2010).

L. Karpa, G. Nikoghosyan, F. Vewinger, M. Fleischhauer, and M. Weitz, Frequency matching in light-storage spectroscopy of atomic raman transitions, Phys. Rev. Lett. 103, 093601 (2009).

T. L. Dao, C. Kollath, I. Carusotto, and M. Köhl, All-optical pump-and-probe detection of two-time correlations in a Fermi gas, Phys. Rev. A 81, 043626 (2010).

F. Beil, M. Buschbeck, G. Heinze, and T. Halfmann, Light storage in a doped solid enhanced by feedback-controlled pulse shaping, Phys. Rev. A 81, 053801 (2010).

A. W. Elshaari, A. Aboketaf, and S. F. Preble, Controlled storage of light in silicon cavities, Opt. Express 18, 3014 (2010).

V. Fiore, Y. Yang, M. C. Kuzyk, R. Barbour, L. Tian, H. Wang, Storing optical information as a mechanical excitation in a silica optomechanical resonator, Phys. Rev. Lett. 107, 133601 (2011).

M. Lobino, C. Kupchak, E. Figueroa, and A. I. Lvovsky, Memory for light as a quantum process, Phys. Rev. Lett. 102, 203601 (2009).

M. Fleischhauer, and M. D. Lukin, Quantum memory for photons: Dark-state polaritons, Phys. Rev. A 65, 022314 (2002).

A. Peng, M. Johnsson, W. P. Bowen, P. K. Lam, H. A. Bachor, and J. J. Hope, Squeezing and entanglement delay using slow light, Phys. Rev. A 71, 033809 (2005).

M. Orrit, Photons pushed together, Nature 460, 42 (2009).

D. A. B. Miller, Are optical transistors the logical next step? Nat. Photon. 4, 3 (2010).

J. Tominaga, C. Mihalcea, D. Büchel, H. Fukuda, T. Nakano, N. Atoda, H. Fuji, and T. Kikukawa, High-speed low-power photonic transistor devices based on optically-controlled gain or absorption to affect optical interference, Appl. Phys. Lett. 78, 2417 (2001).

L. Robinson, Quantum wells for optical logic, Science 225, 822 (1984).

H. Schmidt and R. J. Ram, All-optical wavelength converter and switch based on electromagnetically induced transparency, Appl. Phys. Lett. 76, 3173 (2000).

C. H. R. Ooi, Controlling irreversibility and directionality of light via atomic motion: optical transistor and quantum velocimeter, New J. Phys. 10, 123024 (2008).

P. Bermel, A. Rodriguez, S. G. Johnson, J. D. Joannopoulos, and M. Soljačić, Single-photon all-optical switching using waveguide-cavity quantum electrodynamics, Phys. Rev. A 74, 043818 (2006).

N. Moll, R. Harbers, R. F. Mahrt, and G. L. Bona, Integrated all-optical switch in a cross-waveguide geometry, Appl. Phys. Lett. 88, 171104 (2006).

S. Kim, Z. Wang, D. J. Hagan, E. W. Van Stryland, A. Koyakov, F. Lederer, and G. Assanto, Phase-insensitive all-optical transistors based on second-order nonlinearities, IEEE J. Quantum Electron. 34, 666 (1998).

J. Hwang, M. Pototschnig, R. Lettow, G. Zumofen, A. Renn, S. Götzinger, and V. A. Sandoghdar, Single-molecule optical transistor, Nature 460, 76 (2009).

K. Asakawa, Y. Sugimoto, Y. Watanabe, N. Ozaki, A. Mizutani, and Y. Takata, Photonic crytal and quantum dot technologies for all-optical switch and logic device, New. J. Phys. 8, 208 (2006).

M. F. Yanik, S. H. Fan, and Marin Soljačić, High-contrast all-optical bistable switching in photonic crystal microcavities, Appl. Phys. Lett. 83, 2739 (2003).

M. Freitag, Optical and thermal properties of graphene field-effect transistors, Phys. Status Solidi(B) 247, 2895 (2010).

D. L. Andrews, D. S. Bradshaw, Optical transistor action by nonlinear coupling of stimulated emission and coherent scattering, Proc. SPIE 7797, 77970L (2010).

S. Singh, G. A. Phelps, D. S. Goldbaum, E. M. Wright, P. Meystre, All-optical optomechanics: An optical spring mirror, Phys. Rev. Lett. 105, 213602 (2010).

J. Hofer, A. Schliesser, T. J. Kippenberg, Cavity optomechanics with ultrahigh-Q crystalline microresonators, Phys. Rev. A 82, 031804 (2010).

Chapter 4

Cavity Optomechanical System with Bose-Einstein Condensate

The coherent interaction between matter and a single mode of light is a fundamental theme in cavity quantum electrodynamics, which provides a useful platform for developing concepts in light propagation properties [Brennecke (2007); Zhang (2012); Huang (2012)]. In order to obtain a strong coupling regime, experimenters always using high-quality resonators, where atoms coherently exchange a photon before dissipation sets in [Colombe (2007)] with a single light-field. This has led to fundamental studies with optical resonators [Asjad (2011); Teufel (2011)]. There are challenges in quantum information processing and quantum state engineering, which makes researchers pay too much attention on laser cooling and quantum state engineering [Brennecke (2008); Szirmai (2010); Bhattacherjee (2010); Steinke (2011); Jing (2010); Zhao (2007)]. Confining a dilute gas of bosons in a potential, the weakly interacting ones can be treated as a Bose-Einstein condensate matter, when cooling the temperature to very near absolute zero. In this case, the lowest quantum state of the external potential is occupied by a large fraction of the bosons, which leads to the quantum optical effects in macroscopic scale.

In order to achieve the strong coupling regime, here we use a Bose-Einstein condensate (BEC) coupled to an ultrahigh-finesse optical cavity and present its light propagation properties, under the radiation of two optical fields. This is a conceptually new regime of generalized optomechanical system, in which the collective motion of the BEC can be made an analogy to a mechanical resonator in cavity optomechanical system. Considering a pure BEC, other motional excitations can be safely disregarded and so a situation analogous to optomechanics can be realized, not with a movable mirror but rather with the collective motion of an ensemble of atoms. Here, on one hand, the cavity field will affect the motion degrees of the freedom of the BEC through the exchange of momentum between the cavity field and the BEC. On the other hand, the collective density excitation of the BEC offers feedback on the cavity field by the dependence of the optical path length on the atomic density distribution within the spatially periodic cavity mode structure. The aim of the present chapter is to discuss light propagation properties (slow light [Zhu (2011a)]) as well as quantum optical devices (quantum optical transistor [Zhu (2011b)], single photon router [Zhu (2012c)], and nonlinear Kerr switch [Zhu (2011d)]) in a coupled BEC and optical cavity, especially those close to the region of parameter settings where the optomechanical simplification can be harvested.

(a) Cavity optomechanical system with BEC

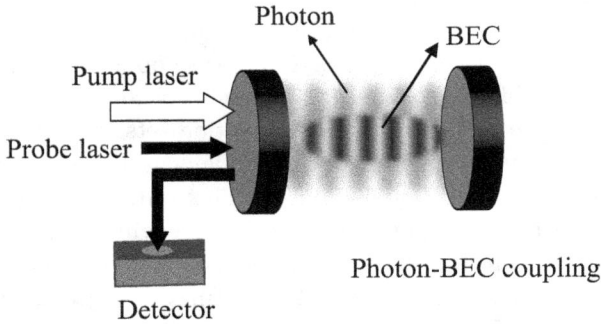

(b) Cavity optomechanical system

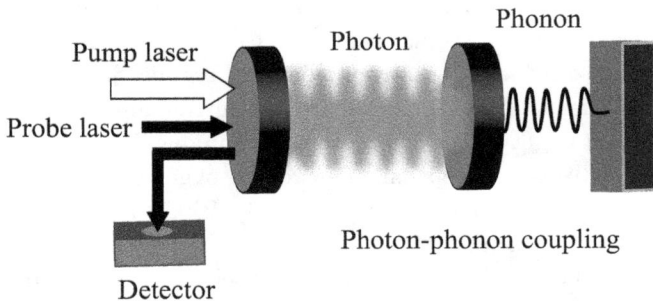

Fig. 4.1   The comparison between the cavity optomechanical system and the cavity optomechanical system with BEC, where the collective motion of the BEC in (a) is analogous with the mechanical resonator in (b).

We study a coupled BEC-cavity system consisting of an optical cavity with a BEC trapped in it, where the collective motion of the BEC can serve as a mechanical oscillator coupled to the cavity filed. Figure 4.1 shows the comparison between the cavity optomechanical system and the cavity optomechanical system with BEC, where the photon-BEC coupling showing in Fig.4.1(a) is in analogy with the photon-phonon coupling in cavity optomechanical system.

Recently, Brennecke *et al.* [Brennecke (2008)] realized the strong coupling of the collective oscillation of the BEC to the quantized cavity field with all atoms in the same motional quantum state, where the atom-atom interactions are neglected. They demonstrated the bistable behavior of BEC-cavity system, and gave a a striking contrast between cavity optomechanical sytem and BEC-cavity system. They proved that the only difference between these two systems is that the mechanical oscillator in BEC-cavity system does not base on the presence of an external harmonic potential. In their paper, the macroscopically occupied zero-momentum state is coupled to the symmetric superposition of the $\pm 2\hbar k$ momentum state via absorption and stimulated emission of cavity photons (k is the wave vector of the cavity photons). The physical process can be well described by the Bogoliubov mode oscillating at the frequency $\omega_m = (2\hbar k)^2/2m\hbar = 2\hbar k^2/m = 4\omega_{\rm rec}$ (m is the mass of the atom and $\omega_{\rm rec}$ is the recoil frequency).

Furthermore Murch's group [Murch (2008)] also studied the BEC-cavity system and demonstrated that it is not necessary to have all atoms coupled equally to the cavity to realize strong coupling. When the atoms are confined at many locations within a one dimensional optical cavity and interact with the cavity mode at the position dependent coupling rate, they demonstrated that the strong coupling can be realized with low intracavity photon numbers. In this case, a more significant like-EIT phenomenon and a greater alteration of the transmitted probe beam can be achieved in this strongly coupled system, which will be discussed in the following.

We take the model in Brennecke [Brennecke (2008)]. The system considered in this chapter is shown in Fig.4.1(a), where a BEC of N ^{87}Rb atoms is trapped within an optical ultrahigh-finesse Fabry-Perot cavity by a crossed beam dipole trap. The Fabry-Perot cavity has been introduced in Chapter 1. We aim for a strong pump field with frequency $\omega_{\rm pu}$ and a weak probe field with frequency $\omega_{\rm pr}$ along the cavity axis. The probe photons transmitted from the cavity is monitored by a detector. We consider the weak atom-atom interactions and a shallow external trapping potential. In a rotating frame at a driving field frequency ω_{pu}, the Hamiltonian of the BEC-cavity system can be given as follows [Brennecke (2008); Paternostro (2010); Gardiner (2004)]:

$$H = H_1 + H_2 + H_3 + H_4 \tag{4.1}$$

$$H_1 = \hbar\omega_m b^\dagger b, \tag{4.2}$$

$$H_2 = \hbar\Delta_c a^\dagger a, \tag{4.3}$$

$$H_3 = \hbar g(b^\dagger + b)a^\dagger a, \tag{4.4}$$

$$H_4 = -i\hbar E_{pu}(a - a^\dagger) - i\hbar E_{pr}(ae^{i\delta t} - a^\dagger e^{-i\delta t}), \tag{4.5}$$

where

$$\omega_m = 4\omega_{rec}, \tag{4.6}$$

$$\Delta_c = \omega_c' - \omega_{pu}, \tag{4.7}$$

$$\omega_c' = \omega_c + \frac{1}{2}U_0 N, \tag{4.8}$$

$$U_0 = \frac{g^2}{\Delta_a}, \Delta_a = \omega_{pu} - \omega_a, \tag{4.9}$$

$$g = \frac{U_0}{2}\sqrt{\frac{N}{2}}, \tag{4.10}$$

$$\delta = \omega_{pr} - \omega_{pu}, \tag{4.11}$$

$$E_{pu} = \sqrt{2P_{pu}\kappa/\hbar\omega_{pu}}, \tag{4.12}$$

$$E_{pr} = \sqrt{2P_{pr}\kappa/\hbar\omega_{pr}}, \tag{4.13}$$

where H_1 is the energy of Bogoliubov mode of the collective oscillation of the BEC (ω_m and b ($b^\dagger$) denote the oscillation frequency and the annihilation (creation) operator of the Bogoliubov mode, respectively). H_2 gives the energy of cavity mode, where ω_c and a ($a^\dagger$) denote the oscillation frequency and the annihilation (creation) operator of the cavity mode, respectively, and $\frac{1}{2}U_0 N$ is the frequency shift of the empty cavity resonance induced by the BEC. N is the number of the condensed atoms, g is the maximum coupling strength between a single atom and a single polarized intracavity photon, and Δ_a is the detuning between the pump laser frequency and atomic D_2 line transition frequency. H_3 describes the coupling between the BEC and the cavity. H_4 represents the classical light inputs with the pump laser power P_{pu} and probe laser power P_{pr}, respectively, and κ is the decay rate of the cavity amplitude.

Here we use the Heisenberg equation of motion to solve the Hamiltonian of BEC-cavity system. According to the communication relations $[a, a^\dagger] = 1$ and $[b, b^\dagger] = 1$, in the presence of the coupling field, we deal with the mean response of the system to the probe field, the equations are written as

$$\frac{d\langle a\rangle}{dt} = -(i\Delta_c + \kappa)\langle a\rangle - i\sqrt{2}g\langle X\rangle\langle a\rangle + E_{pu} + E_{pr}e^{-i\delta t}, \quad (4.14)$$

$$\frac{d^2\langle X\rangle}{dt^2} + \gamma_m\frac{d\langle X\rangle}{dt} + \omega_m^2\langle X\rangle = -\omega_m g\sqrt{2}\langle a^\dagger\rangle\langle a\rangle, \quad (4.15)$$

where $X = (b + b^\dagger)/\sqrt{2}$, and $\langle a\rangle$, $\langle a^\dagger\rangle$ and $\langle X\rangle$ are the expectation values of operators a, $a^\dagger$ and X, respectively. γ_m is the damping rate of the Bogoliubov mode of the collective oscillation of BEC. To solve these equations, we make the ansatz as follows $\langle a(t)\rangle = a_0 + a_+e^{-i\delta t} + a_-e^{i\delta t}$, $\langle X(t)\rangle = X_0 + X_+e^{-i\delta t} + X_-e^{i\delta t}$.

By substituting these approximation into (4.14) and (4.15), respectively, and upon working to the lowest order in E_{pr} but to all orders in E_{pu}, we can get

$$a_+ = E_{pr}\left[\frac{(\kappa - i\delta) - i(\Delta_c + iC)}{(\kappa - i\delta)^2 + (\Delta_c + iC)^2 - D}\right], \quad (4.16)$$

and

$$\omega_0[\kappa^2 + (\Delta_c - \frac{2g^2}{\omega_m}\omega_0)^2] = E_{pu}^2,$$

with $C = iA\omega_m\omega_0(1 + B)$, $D = A^2B^2\omega_m^2\omega_0^2$, $A = 2g^2/\omega_m^2$, $B = \omega_m^2/(\omega_m^2 - i\delta\gamma_m - \delta^2)$ and $\omega_0 = |a_0|^2$.

Using input-output relation [Gardiner (2004); Weis (2010)], the optical property of the output field can be expressed as, $a_{out}(t) = a_{in}(t) - \sqrt{2\kappa}a(t)$, where a_{in} and a_{out} are the input and output operator, respectively.

The expectation value of the output field can be written as

$$\langle a_{out}(t)\rangle = (E_{pu}/\sqrt{2\kappa} - \sqrt{2\kappa}a_0)e^{-i\omega_{pu}t}$$
$$+ (E_{pr}/\sqrt{2\kappa} - \sqrt{2\kappa}a_+)e^{-i(\omega_{pu}+\delta)t}$$
$$- \sqrt{2\kappa}a_-e^{-i(\omega_{pu}-\delta)t}. \quad (4.17)$$

The transmission of the probe beam, defined as the ratio of the output and input field amplitudes at the probe frequency is then given by [Weis (2010); Safavi-Naeini (2011)]

$$t(\omega_{pr}) = \frac{E_{pr}/\sqrt{2\kappa} - \sqrt{2\kappa}a_+}{E_{pr}/\sqrt{2\kappa}} = 1 - 2\kappa a_+/E_{pr}, \quad (4.18)$$

and the transmission group delay can be expressed as [Weis (2010)]

$$T_t = \frac{d\phi_t(\omega_{\mathrm{pr}})}{d\omega_{\mathrm{pr}}}\bigg|_{\omega_{\mathrm{pr}}=\omega_c'} = \frac{d\{\arg[t(\omega_{\mathrm{pr}})]\}}{d\omega_{\mathrm{pr}}}\bigg|_{\omega_{\mathrm{pr}}=\omega_c'}. \tag{4.19}$$

4.1 Slow light

We choose the realistic parameters of the BEC-cavity system in [Brennecke (2008)], where $N = 1.2 \times 10^5$, $g = 2\pi \times 10.9\mathrm{MHz}$, $\kappa = 2\pi \times 1.3\mathrm{MHz}$, $\Delta_a = 2\pi \times 32\mathrm{GHz}$, $\gamma_m = 2\pi \times 0.4\mathrm{kHz}$ and $\omega_{\mathrm{rec}} = 2\pi \times 3.8\mathrm{kHz}$.

Figure 4.2 shows the probe transmission spectrum $|t|^2$. It is shown that in presence of pump laser, the transmission spectrum of the probe beam displays a significant transparency window at the resonant region ($\Delta'_{\mathrm{prc}} = 0$). The physical phenomenon is similar to electromagnetically induced transparency (EIT), which has been discussed in Chapter 3. Here, the collective

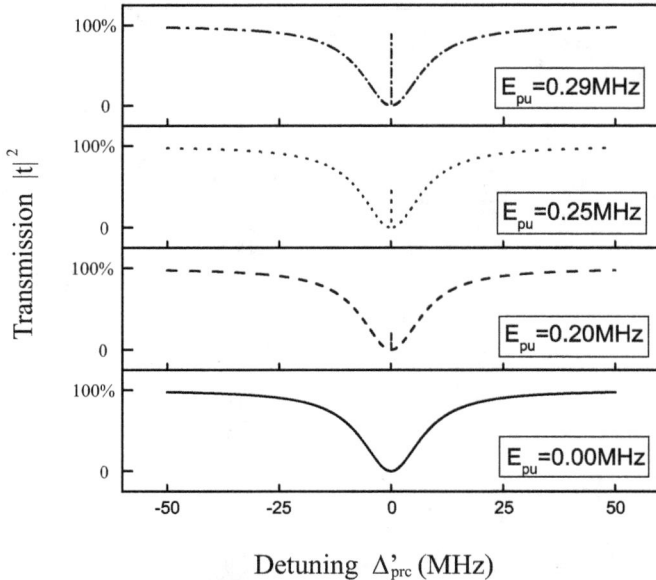

Fig. 4.2   The transmission $|t|^2$ of the probe beam as a function of the effective probe-cavity detuning $\Delta'_{\mathrm{prc}} = \omega_{\mathrm{pr}} - \omega_c'$ with different E_{pu}. The parameters used are $\Delta_c = \omega_m$, $N = 1.2 \times 10^5$, $g = 2\pi \times 10.9\mathrm{MHz}$, $\kappa = 2\pi \times 1.3\mathrm{MHz}$, $\Delta_a = 2\pi \times 32\mathrm{GHz}$, $\gamma_m = 2\pi \times 0.4\mathrm{kHz}$ and $\omega_{\mathrm{rec}} = 2\pi \times 3.8\mathrm{kHz}$.

motion of the BEC can be made an analogy to a mechanical resonator with resonance frequency ω_m. The simultaneous presence of pump and probe laser fields induces a radiation-pressure force oscillating at the beat frequency $\delta = \omega_{pr} - \omega_{pu}$, which drives the collective motion of the BEC near its resonance frequency. As shown in Fig.4.1, the photon-BEC coupling is in analogy with the photon-phonon coupling via radiation pressure. If the beat frequency δ is close to the resonance frequency ω_m of the BEC, the mechanical mode starts to oscillate coherently, which will result in Stokes ($\omega_s = \omega_{pu} - \omega_m$) and anti-Stokes ($\omega_{as} = \omega_{pu} + \omega_m$) scattering of light from the intracavity field. For the near-resonant probe laser($\Delta'_{prc} = 0$), the probe field can interfere with the anti-Stokes field and as a result the probe spectrum can be modified significantly [Teufel (2011)].

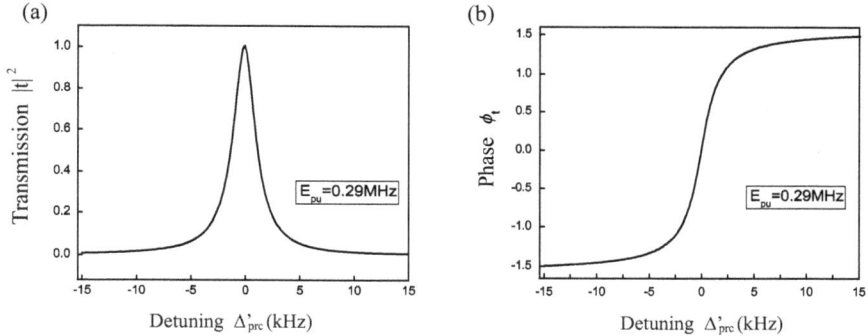

Fig. 4.3   (a) The magnified transparency window as a function of Δ'_{prc}. Other parameters are $N = 1.2 \times 10^5$, $g = 2\pi \times 10.9\text{MHz}$, $E_{pu} = 0.29\text{MHz}$, $\kappa = 2\pi \times 1.3\text{MHz}$, $\Delta_a = 2\pi \times 32\text{GHz}$, $\gamma_m = 2\pi \times 0.4\text{kHz}$ and $\omega_{rec} = 2\pi \times 3.8\text{kHz}$. (b) The phase of the transmitted probe beam as a function of Δ'_{prc}.

Figure 4.3 shows the magnified transparency window and the phase of the transmitted probe field, respectively. As shown in Fig.4.3(a), on resonance, the transmission coefficient is one, but far off resonance it rapidly reduces to zero. With the increase of Δ'_{prc}, Fig.4.3(b) shows that the phase of the transmitted probe beam suffers a sharp enhancement, which reveals that the transmitted probe beam can be significantly delayed by the BEC-cavity system.

By tuning the pump laser to the lower motional sideband of the cavity, Fig.4.4 shows that in this coupled BEC-cavity system, the slow light can be realized. With the decrease of E_{pu} the propagation velocity of the probe

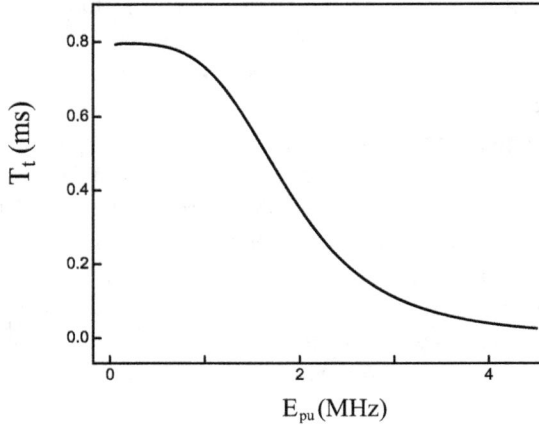

Fig. 4.4   The transmission group delay T_t as a function of E_{pu}. Other parameters are $\Delta_c = \omega_m$, $N = 1.2 \times 10^5$, $g = 2\pi \times 10.9\text{MHz}$, $\kappa = 2\pi \times 1.3\text{MHz}$, $\Delta_a = 2\pi \times 32\text{GHz}$, $\gamma_m = 2\pi \times 0.4\text{kHz}$ and $\omega_{rec} = 2\pi \times 3.8\text{kHz}$.

pulse can be slow down quickly, i.e., $E_{pu} = 0.25\text{MHz}$, the peak of the probe pulse can be delayed by 0.8 ms.

4.2   All-optical transistor

Nowadays, the need for faster and more powerful computer and internet connection gets high demand [Miller (2010)]. However, conventional information precessing using electrons as signal carrier limits the performance of computers, because they are forced to be operated at ever-higher frequencies, power dissipation and consequent hardware heating [Joannopoulos (2004)]. Gradually, photons become signal carrier in powering processors rather than electrons, because of the much less heat and the much higher transfer rates within the device [Orrit (2009)]. Consequently, optical transistors are an attractive sought goal because they could form the basis of quantum computers and optical computers that use photons instead of electrons as signal carriers [Gibbs (1985); O'Brien (2007); Bouwmeester (2000)]. Under the radiation of two optical fields, an all-optical transistor can be realized if the propagation of optical 'signal' field can be effectively controlled by another optical 'gate' field [Zhu (2011b); Chang (2007); Hwang (2009)].

However, the weakness of the photon-photon interaction leads to the challenging of an all-optical transistor. Recently, by exploiting the strong coupling between individual optical emitter and surface plasmons confined to a conducting nanowire, Chang *et al.* [Chang (2007)] have put forward single-photon transistor in the strong regime of nonlinear interactions between single photons. Furthermore, Hwang *et al.* [Hwang (2009)] have proposed an all-opitcal transistor in a single dye molecule, where a tightly focused laser beam (the 'signal' field) can coherently be attenuated or amplified, via effectively regulating the second 'gate' field.

In the above section, we demonstrate that a strong coupling regime can be reached just by regulating the pump power. Here we propose an all-optical transistor in a coupled BEC-cavity GOS at the blue detuning. In the strong coupling optomechanical system, the motion of the oscillator can easily modulate the optical path length and thus the frequency of the cavity, creating sidebands above (anti-Stokes sideband) and below (Stokes sideband) the drive frequency. For a near-resonant probe laser, due to its interference with the induced sidebands, the propagation of the probe beam can be strongly altered by the optomechanical system, which has been demonstrated theoretically by Agarwal and Huang [Agarwal (2010)] and experimentally by Weis *et al.* [Weis (2010)], Safavi-Naeini *et al.* [Safavi-Naeini (2011)], and Teufel *et al.* [Teufel (2011)], respectively.

Figure 4.5 shows the probe transmission spectrum $|t|^2$ as a function of Δ'_{prc}, where $\Delta'_{\mathrm{prc}} = \omega_{\mathrm{pr}} - \omega_c - \frac{1}{2}NU_0$. This figure indicates that the transmission spectrum of the probe laser can be modulated by the pump laser effectively. Turning off or on the pump laser correspond to the transmitted and reflected probe laser respectively, i.e., if $P_{\mathrm{pu}} = 0$ (the solid curve), the the probe laser can not transmit the cavity; if $P_{\mathrm{pu}} \neq 0$ and fix the pump laser on the blue cavity-pump detuning ($\Delta_c = -\omega_m$), significant transmissions of the probe laser can be obtained at $\Delta'_{\mathrm{prc}} = 0$. Particularly, when $P_{\mathrm{pu}} = 1.3\mathrm{fW}$, the probe laser is totally transmit the cavity. As the power of the pump laser increases further, the transmission of the probe laser can be amplified significantly. For example, when $P_{\mathrm{pu}} = 1.7\mathrm{fW}$, the probe transmission is amplified to 175%.

At the beat frequency δ, the radiation pressure force drives the collective motion of the BEC near its resonant frequency under resonant cavity-pump detuning ($\Delta_c = \pm\omega_m$). If the beat frequency δ is close to the resonance frequency ω_m of the BEC, the oscillated mechanical resonator will produce Stokes ($\omega_s = \omega_{\mathrm{pu}} - \omega_m$) and anti-Stokes ($\omega_{as} = \omega_{\mathrm{pu}} + \omega_m$) scattering of light from the strong intracavity field, as discussed in last chapter. For the red detuning (slow light effect), i.e., $\Delta_c = \omega_m$, the anti-Stokes field interferes with the probe field for the near-resonant probe($\Delta'_{\mathrm{prc}} = 0$) and the

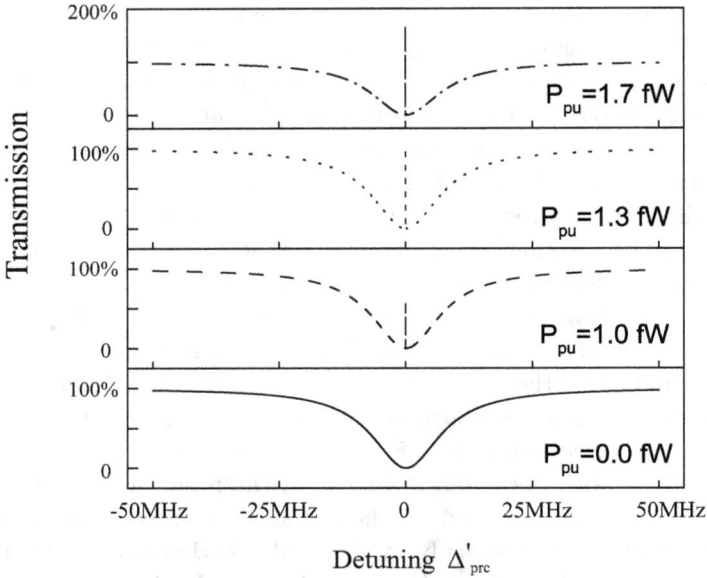

Fig. 4.5   The transmission $|t|^2$ of the probe beam as a function of the shifted probe-cavity detuning Δ'_{prc} with different pump powers. The parameters used are $\Delta_c = -\omega_m$, $N = 1.2 \times 10^5$, $g = 2\pi \times 10.9\text{MHz}$, $\kappa = 2\pi \times 1.3\text{MHz}$, $\Delta_a = 2\pi \times 32\text{GHz}$, $\gamma_m = 2\pi \times 0.4\text{kHz}$ and $\omega_{\text{rec}} = 2\pi \times 3.8\text{kHz}$.

effective interaction Hamiltonian between the cavity modes and mechanical phonon modes has the form of the well-known beam-splitter one in quantum optics [Fiore (2011)]. However, for the blue detuning, i.e., $\Delta_c = -\omega_m$, it is the Stokes field that interferes with the near-resonant probe field and thus modifies the probe spectrum.

Figure 4.6 shows the magnified transparency window of the transmit-ted probe beam as a function of Δ'_{prc} with a blue cavity-pump detuning ($\Delta_c = -\omega_m$). It is clear that a transparency window of several kHz is achieved around the resonance ($\Delta'_{\text{prc}} = 0$). Increasing the pump power, the transmission coefficient of probe laser is largely magnified when $\Delta_c = -\omega_m$.

It is should be noticed that all the magnified curves we observed are in the condition of blue cavity-pump detuning, i.e., $\Delta_c = -\omega_m$. Figure 4.7 plots the relationship between the probe transmission $|t|^2$ and the pump power P_{pu}, which displays that with the enhancement of the pump power the probe transmission increases greatly at first and finally reaches satura-tion at a critical pump power. Moreover, it can be seen that in the interval

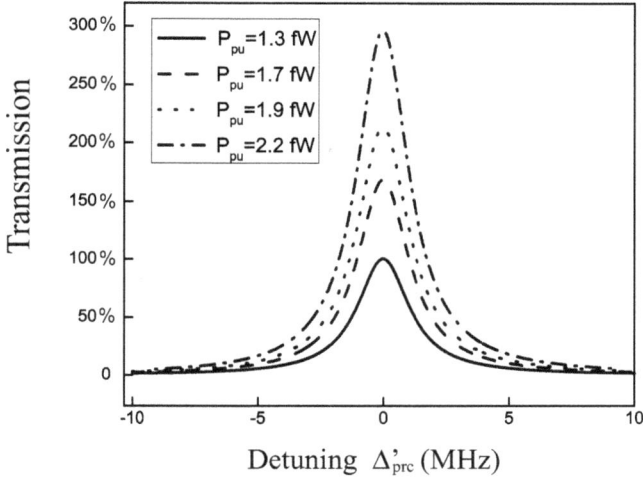

Fig. 4.6 The magnified transparency window as a function of Δ'_{prc} with different pump powers. Other parameters are $\Delta_c = -\omega_m$, $N = 1.2 \times 10^5$, $g = 2\pi \times 10.9\text{MHz}$, $\kappa = 2\pi \times 1.3\text{MHz}$, $\Delta_a = 2\pi \times 32\text{GHz}$, $\gamma_m = 2\pi \times 0.4\text{kHz}$ and $\omega_{\text{rec}} = 2\pi \times 3.8\text{kHz}$.

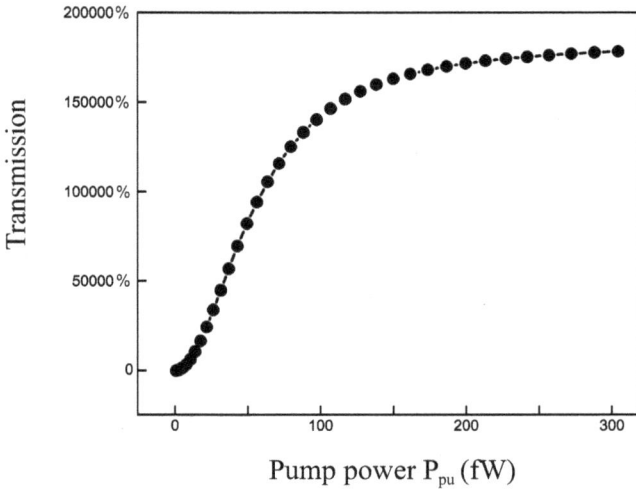

Fig. 4.7 The probe transmission spectrum as a function of the pump power P_{pu} with $\Delta_c = -\omega_m$. Other parameters are $N = 1.2 \times 10^5$, $g = 2\pi \times 10.9\text{MHz}$, $\kappa = 2\pi \times 1.3\text{MHz}$, $\Delta_a = 2\pi \times 32\text{GHz}$, $\gamma_m = 2\pi \times 0.4\text{kHz}$ and $\omega_{\text{rec}} = 2\pi \times 3.8\text{kHz}$.

of $P_{pu} = 19 \sim 52$fW, the system presents a greater switching efficiency. In this case, the maximum amplification of the transmitted probe laser is about 200% when $P_{pu} = 0.1$fW. And if $P_{pu} > 52$fW, the switching efficiency starts to reduce and then approaches saturation at $P_{pu} \approx 200$fW.

Due to the probe transmission can be effectively modulated by the pump beam, we propose a scheme for realizing an all-optical transistor based on the BEC-cavity GOS. Here, the pump laser acts as a 'gate' beam, which controls the attenuation and amplification of the probe laser in blue detuning sideband, while fixing the $\Delta'_{prc} = 0$. In the absence of the pump beam, the probe beam can fully be attenuated. However, as the pump beam is presented with an ideal power ($P_{pu} = 1.3$fW), the cavity becomes completely transparent to the probe beam.

4.3   Single photon router

Recently, single photon router as a rudimentary quantum node of a quantum network has attracted many researcher's attention [Kimble (2008); Aoki (2009); Hoi (2011); Agarwal (2012)]. The first single photon router was realized by Hoi *et al.* [Hoi (2011)] in the microwave regime, where a superconducting transmon qubit is coupled to a superconducting transmission line. They experimentally demonstrated how to choose the output signal from different channels via controlling another pump field on or off. Recently, Agarwal *et al.* [Agarwal (2012)] have theoretically proposed a single photon router in cavity optomechanical system, where a single photon probe field could also be routed effectively.

With the rapid advancement of the nanotechnology, the BEC can be trapped on a small scale [Wang (2005); Fortágh (2007)] and thus a robust miniature device can easily be implemented [Colombe (2007); Treutlein (2007)]. It is well known that for quantum information networks, controlling the output signal of the optical routers or switches with a low power coupling beam is very important in order to use the routers or switches as cascaded classical or quantum computational elements [Dawes (2005); Keyes (1970)]. Therefore, the coupled BEC-cavity system will be a promising candidate for realizing a single photon router. In the above sections, we demonstrate that when applying a pump laser as well as a weak probe one to drive the cavity, the coupling between the collective oscillation of the BEC and the intracavity field can be controlled effectively by the pump field and the probe spectrum can be modulated simultaneously. In this section, we propose a novel scheme to realize a single photon router in a coupled BEC-cavity system. We analyze the reflection and the transmission spectrum of the probe beam with and without the pump beam, respectively. The obtained results show that the route of the probe photon can effectively be controlled by the pump beam with an ultra-low power.

In above section, the optical transistor is working on the blue detuning sideband ($\Delta'_c = -\omega_m$). However, the single photon router proposed here is operated in the red sideband ($\Delta'_c = \omega_m$).

In the absence of pump field, Fig.4.8(a) shows the reflectance $|r|^2 = 1 - |t|^2$ [Zhu (2012c)] and transmittance $|t|^2$ of the probe beam as a function of the effective probe-cavity detuning $\Delta'_{\text{prc}} = \omega_{\text{pr}} - \omega'_c$. At the resonant region ($\Delta'_{\text{prc}} = 0$), the probe beam can completely transmit the cavity from the right port but its reflection is totally suppressed. However, while applying an appropriate pump laser to drive the cavity, the reflection spectrum and the transmission spectrum exhibit a completely different feature. Figure 4.8(b) presents the reflectance and transmittance of the probe beam near resonance as a function of Δ'_{prc}. Obviously, on resonance $|r|^2 = 1$ but $|t|^2 = 0$, which means when the cavity is driven by an appropriate pump field, the resonant probe field can be reflected completely while its transmission is totally suppressed. It should be noted that the reflection peak in Fig.4.8(b) is approximately four orders of magnitude narrower than the transmission peak in Fig.4.8(a), due to the damping rate of the collective oscillation of BEC is much smaller than the decay rate of the cavity field.

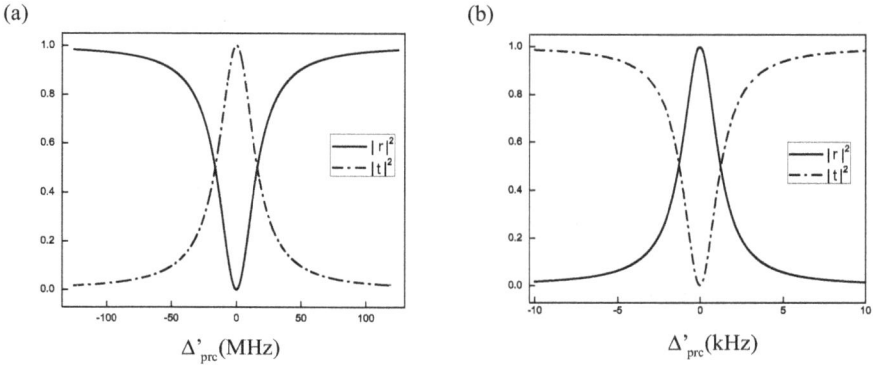

(a)   (b)

Δ'_{prc}(MHz) Δ'_{prc}(kHz)

Fig. 4.8   The reflectance $1 - |t|^2$ and transmittance $|t|^2$ of the probe beam respectively as a function of Δ'_{prc} with $\Delta'_c = \omega_m$ for (a) $P_{\text{pu}} = 0$W and (b) $P_{\text{pu}} = 1.05 \times 10^{-14}$W. Other parameters are $N = 1.2 \times 10^5$, $g = 2\pi \times 10.9$MHz, $\lambda_{\text{pu}} = 780$nm, $\kappa = 2\pi \times 1.3$MHz, $\Delta_a = 2\pi \times 32$GHz, $\gamma_m = 2\pi \times 0.4$kHz and $\omega_{\text{rec}} = 2\pi \times 3.8$kHz.

Due to the switch behavior of the BEC-cavity system discussed above, we propose a single photon router, where the transmission of the probe beam can be controlled effectively by the pump beam. First, a weak probe signal is input from the left port of the cavity, and then we apply a second red detuning $(\Delta'_c = \omega_m)$ pump laser to control the propagation of the probe photon. The output probe single from the right port of cavity can be transmitted and reflected completely, while turning off and on the pump field, respectively. Here, the single photon router based on BEC-cavity not only with ultra-low power, but also operated in all-optical domain, rather than the microwave regime, which is required for a practical router used in the quantum information networks [Dawes (2005); Keyes (1970)]. As shown in Fig.4.8(b), the pump power of this single photon router is as low as 1.05×10^{-14}W.

The Bose-Einstein condensate plays an important role for routing the probe photon. There are two situations: (1)pump off: the effective coupling strength $(G = g\sqrt{\omega_m})$ between the collective oscillation of the BEC and the cavity field is zero, which means the collective oscillation of the BEC has no effect on the propagation of the probe photon; (2)pump on: a radiation-pressure force oscillating at the beat frequency δ, which drives the collective motion of the BEC near its resonance frequency ω_m. If $\delta \approx \omega_m$, the mechanical mode starts to oscillate coherently, which will result in Stokes $(\omega_s = \omega_{\text{pu}} - \omega_m)$ and anti-Stokes $(\omega_{\text{as}} = \omega_{\text{pu}} + \omega_m)$ scattering of light from the intracavity field [Weis (2010)]. For the near-resonant probe, the probe field can interfere with the anti-Stokes field and thus the probe beam is full reflected. In this case, the BEC acts as a perfectly reflective mirror.

4.4 Nonlinear all-optical Kerr switch

All-optical switch plays an important role in quantum information networks, where one light beam can be fully controlled by another. It operates at a fundamental limit of a small number of photons or sometimes even of one photon per switching event. Controlling the output beam of the optical switches by a weaker switching beam is very important to use the switches as cascaded classical or quantum computational elements. However, for most conventional materials working at low power of the switch beam, only a very small nonlinear response can be obtained, which can be significantly outweighed by the linear optical absorption. To overcome these obstacles, based on electromagnetically induced transparency (EIT), Schmidt and Imamoğlu [Schmidt (1996)] implemented a significant enhancement of the third-order optical susceptibility in a four-level system, while completely suppressing the linear optical susceptibility. Since this pioneering work, some other schemes are proposed to realized all-optical switches [Dawes (2005); Tanabe (2005); Zhang (2007); Bajcsy (2009); Scheuer (2010); Qin (2010); Wei (2010)].

By trapping a few hundred cold atoms inside the hollow core of a photonic-crystal fiber, Bajcsy *et al.* [Bajcsy (2009)] demonstrated a fiber-optical switch with a four-level system. The transmission of the probe beam can be effectively controlled by modulating the switching field, which served as an efficient nonlinear optical switch. Furthermore, Scheuer *et al.* [Scheuer (2010)] reported an all-optical switch by coupling two microrings with two waveguides. By aiming the signal laser and pump laser to the lower and upper waveguides and turning the pump beam on and off, they can switch the signal spectral properties between a "dark state" (the off-state) and electromagnetically induced transparency (the on-state).

For a coupled BEC-cavity GOS, the atoms are in the same quantum state, so the coupling to the cavity mode is identical for all atoms. As a result, even a few photons inside the cavity can trigger a strong coupling strength between the BEC and the cavity field, which has been approached experimentally by Brennecke *et al.* [Brennecke (2008)]. Therefore, a significant nonlinear response in BEC-cavity GOS with a low pump power can be realized. In this section, we present a scheme to realize a nonlinear Kerr switch in a coupled BEC-cavity GOS in all-optical domain. Based on the Bogoliubov two-mode approximation [Brennecke (2008); Nagy (2009)], the collective motion of the BEC can be made an analogy to an oscillation of a moving mirror in the optomechanical system. The cavity field coupled to a collective density excitation of the BEC can result in an extremely large coupling strength even for low pump power [Brennecke (2008)]. And the large nonlinear optical response can be obtained, via modulating the pump laser properly.

The system considered here is shown in Fig.4.1(a). From Eq.(4.14) and Eq.(4.15), we can obtain:

$$a_- = \frac{AB^* \omega_m E_{pr} E_{pu}^2}{(\kappa + i\delta)^2 + (\Delta_c - iC^*)^2 - D^*} \cdot \frac{1}{(i\Delta_c - iA\omega_m\omega_0 + \kappa)^2}. \quad (4.20)$$

To investigate the optical property of the output field for this coupling system, using input-output relation [Weis (2010); Gardiner (1985, 2004)], we can obtain the output field as

$$\langle a_{\text{out}}(t)\rangle = \langle a_{\text{in}}(t)\rangle - \sqrt{2\kappa}\langle a(t)\rangle$$
$$= (E_{\text{pu}}/\sqrt{2\kappa} - \sqrt{2\kappa}a_0) + (E_{\text{pr}}/\sqrt{2\kappa} - \sqrt{2\kappa}a_+)e^{-i\delta t}$$
$$- \sqrt{2\kappa}a_- e^{i\delta t}. \tag{4.21}$$

We can define $\langle a_{\text{out}}(t)\rangle$ as

$$\langle a_{\text{out}}(t)\rangle = a_{\text{out}^0} + a_{\text{out}+}e^{-i\delta t} + a_{\text{out}-}e^{i\delta t}. \tag{4.22}$$

and then

$$a_{\text{out}-} = \frac{E_{\text{pu}}^2 E_{\text{pr}}}{(i\Delta_c - iA\omega_m\omega_0 + \kappa)^2} \cdot \frac{\sqrt{2\kappa}AB^*\omega_m}{D^* - (\kappa + i\delta)^2 - (\Delta_c - C^*)^2}. \tag{4.23}$$

Equation (4.24) shows that the $a_{\text{out}-}$ are proportional to E_{pr} and $E_{\text{pu}}^2 E_{\text{pr}}$, which can be made an analogy to the third-order nonlinear polarization [Boyd (2008)]. The term $a_{\text{out}-}$ exhibits the cross-Kerr nonlinear response. The relative transmitted intensity corresponding to the cross-Kerr response can be defined as

$$T_{\text{out}-} = |\frac{c_{\text{out}-}}{E_{\text{pr}}/\sqrt{2\kappa}}|^2$$
$$= |\frac{2\kappa AB^*\omega_m E_{\text{pu}}^2}{(i\Delta_c - iA\omega_m\omega_0 + \kappa)^2 \cdot [D^* - (\kappa + i\delta)^2 - (\Delta_c - C^*)^2]}|^2. \tag{4.24}$$

Because of a large detuning Δ_a, it should be noted that the spontaneous emission is negligible and the atomic excited state can be eliminated adiabatically [Brennecke (2008); Nagy (2009); Bhattacherjee (2009)], which means, unlike the strong interactions between the single atom and the cavity photons done by Kimble [Kimble (1998)], the cavity-single atom interaction described here does not play any role. It is a collective density excitation of the BEC who acts as a mechanical oscillator and strongly couples to the cavity. The light will affect the motion degrees of the freedom of the BEC through the exchange of momentum between the light and the BEC. Besides, the collective density excitation of the BEC offers feedback on the cavity field by the dependence of the optical path length on the atomic density distribution within the spatially periodic cavity mode structure [Brennecke (2008)].

Figure 4.9 plots the relationship between the mean photon number ω_0 of the intracavity and the cavity-pump detuning Δ_c with different E_{pu}, which shows that there is a turning-point for the bistable behavior of BEC-cavity GOS. At the low pump power, the above two curves of Fig.4.9 exhibit

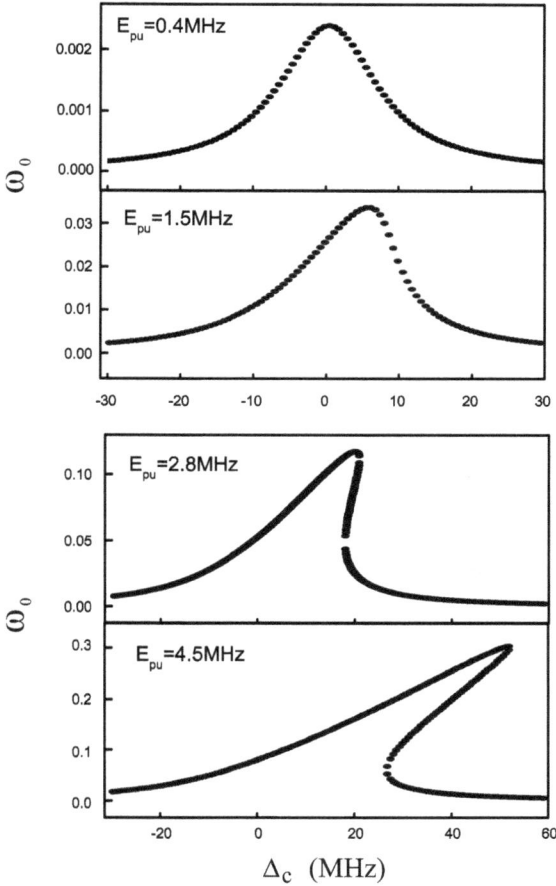

Fig. 4.9 The mean photon number ω_0 of the intracavity as a function of cavity-pump detuning Δ_c. The other parameters used are $N = 1.2 \times 10^5$, $g_0 = 2\pi \times 10.9\text{MHz}$, $\kappa = 2\pi \times 1.3\text{MHz}$, $\Delta_a = 2\pi \times 32\text{GHz}$, $\gamma_m = 2\pi \times 0.4\text{kHz}$ and $\omega_{\text{rec}} = 2\pi \times 3.8\text{kHz}$.

non-bistable feature, while increasing the pump power results in a bistable behavior as shown in the below two curves of Fig.4.9. Figure 4.10 presents a significant transmitted peak accompanying a sharp change of the phase near the resonant region($\Delta'_{\text{prc}} = 0$), where (a) the relative transmitted intensity $T_{\text{out}-}$ and (b) the phase ϕ_T corresponding to the cross-Kerr response. For the near-resonant probe laser($\Delta'_{\text{prc}} = 0$), the probe field can interfere with the anti-Stokes field, which will modified the probe spectrum significantly, and eventually results in the sharp change of probe phase [Teufel (2011)].

(a)

(b)

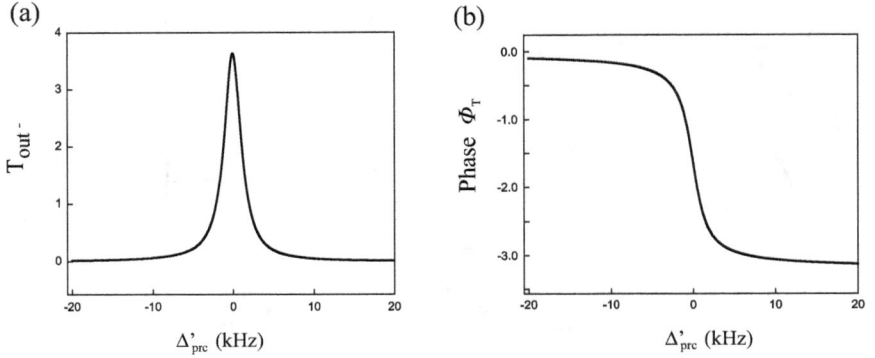

Fig. 4.10 (a) The relative transmitted intensity $T_{\text{out}-}$ and (b) the phase ϕ_T corresponding to the cross-Kerr response as a function of the effective probe-cavity detuning $\Delta'_{\text{prc}} = \omega_{\text{pr}} - \omega'_c$. The other parameters used are $\Delta_c = \omega_m$, $E_{\text{pu}} = 0.4\text{MHz}$, $N = 1.2\times10^5$, $g_0 = 2\pi\times10.9\text{MHz}$, $\kappa = 2\pi\times1.3\text{MHz}$, $\Delta_a = 2\pi\times32\text{GHz}$, $\gamma_m = 2\pi\times0.4\text{kHz}$ and $\omega_{\text{rec}} = 2\pi \times 3.8\text{kHz}$.

To explore the physical mechanism of the Kerr nonlinearity in the coupled BEC-cavity system, the dependence of the relative cross-Kerr transmitted intensity $T_{\text{out}-}$ on the effective coupling strength and pump power are plotted in Fig.4.11. In Fig.4.11(a), an obvious transmitted peak can be obtained at $g = 2.86\text{MHz}$. However, if we do not consider the collective oscillation of BEC and assume $g = 0$, the optical cross-Kerr response disappears immediately (the dotted curve in Fig.4.11(a)). Therefore, we can conclude that the collective oscillation of the BEC plays a vital role for the generation of the Kerr response in this coupled system. Because of the interaction between the collective oscillation mode of the BEC and the cavity field, the momentum and energy can be exchanged between them, which leads to the production of the optical Kerr nonlinearity. Figure 4.11(b) shows that the pump power also generates a significant implication for the cross-Kerr transition, which demonstrates that the magnitude of the optical Kerr response can be modulated by the pump laser efficiently. Turning on and turning off the pump laser correspond to the on-state and off-state of optical Kerr switch, respectively. Increasing the power of the pump laser, the optical transmitted intensity will be enhanced significantly. For example, if the intracavity photon number $n_d = 10^6$, the number of the condensed atoms $N = 1.2 \times 10^5$ and the amplitude of the pump laser $E_{\text{pu}} = 0.4\text{MHz}$ [Brennecke (2008)], the optical Kerr coefficient can be $\approx 2.1 \times 10^{-17}\text{m}^2/\text{V}^2$. However, when turning the pump field off ($E_{\text{pu}} = 0$), this optical Kerr response will disappear immediately.

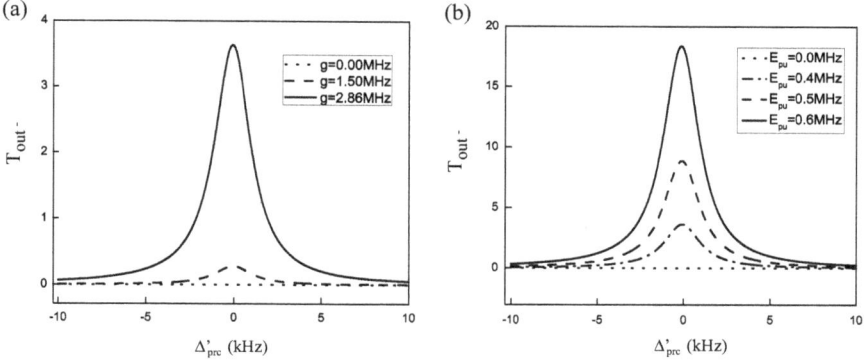

Fig. 4.11   The relative cross-Kerr transmitted intensity $T_{\mathrm{out}-}$ as a function of the effective probe-cavity detuning $\Delta'_{\mathrm{prc}} = \omega_{\mathrm{pr}} - \omega'_c$ with (a) different effective coupling strengths (g) $E_{\mathrm{pu}} = 0.4\mathrm{MHz}$ for three different effective coupling strengths (g) and (b) different E_{pu}. The other parameters used are $\Delta_c = \omega_m$, $N = 1.2 \times 10^5$, $g_0 = 2\pi \times 10.9\mathrm{MHz}$, $\kappa = 2\pi \times 1.3\mathrm{MHz}$, $\Delta_a = 2\pi \times 32\mathrm{GHz}$, $\gamma_m = 2\pi \times 0.4\mathrm{kHz}$ and $\omega_{\mathrm{rec}} = 2\pi \times 3.8\mathrm{kHz}$.

The physical mechanism of the Kerr nonlinearity proposed here is: when driving the optical cavity with a strong pump field, the intracavity photons are excited and generate a strong interaction with collective oscillation mode via enhanced radiation pressure, which will eventually leads to the enhancement of the Kerr optical nonlinearity. On the contrary, in the absence of the pump laser, the interaction between the collective oscillation mode and the cavity mode becomes very weak, and the Kerr optical nonlinearity will disappear. Because the pump laser can effectively modulate the Kerr optical nonlinearity, the BEC-cavity GOS under the radiation of two optical fields can really act as an all-optical Kerr switch.

Next, we show the operation power and the switching time of this optical Kerr switch. Figure 4.12 presents the peak values of the relative cross-Kerr transmitted intensity as a function of pump beam power P_{pu} with $\Delta_c = \omega_m$. Compared with the work done by Dawes *et al.* [Dawes (2005)] where the power of the switching beam is $230 \times 10^{-12}\mathrm{W}$, and the work carried out by Zhang *et al.* [Zhang (2007)] where the optical switch is operated at a lower switching power about $10^{-12}\mathrm{W}$, the optical Kerr switch based on BEC-cavity GOS have a saturated pump power at $\approx 1.5 \times 10^{-13}\mathrm{W}$, which is one order of the magnitude lower than previous works. However, it has been demonstrated that the ultra-low power will prolong the switching, which is determined by the ground state decoherence rate in cavity QED system [Weis (2010)]. In our BEC-cavity system, such a rate is related to

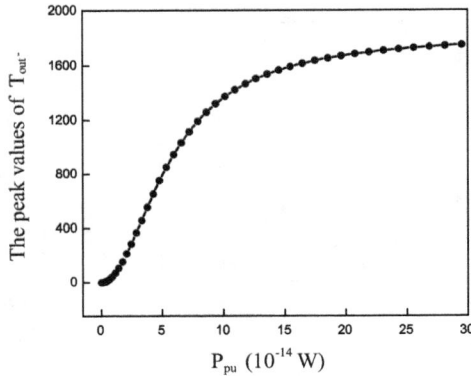

Fig. 4.12   The peak values of the relative cross-Kerr transmitted intensity T_{out-} as a function of the pump power P_{pu}. The other parameters used are the same with Fig.4.11.

the damping rate of the collective oscillation of the BEC, which can be reached at $\approx 1/\gamma_m \sim 4 \times 10^{-4}$s. This is one order of the magnitude faster than the maximum switching time reported by Weis *et al.* [Weis (2010)].

Bibliography

F. Brennecke, T. Donner, S. Ritter, T. Bourdel, M. Köhl, and T. Esslinger, Cavity QED with a Bose-Einstein condensate, Nature 450, 268 (2007).

Y. Zhang, L. Mao, C. Zhang, Mean-field dynamics of spin-orbit coupled Bose-Einstein condensates, Phys. Rev. Lett. 108, 035302 (2012).

J. F. Huang, Q. Ai, Y. Deng, C. P. Sun, F. Nori, Quantum statistics of the collective excitations of an atomic ensemble inside a cavity, Phys. Rev. A 85, 023801 (2012).

Y. Colombe, T. Steinmetz, G. Dubois, F. Linke, D. Hunger, J. Reichel, Strong atom-field coupling for Bose-Einstein condensates in an optical cavity on a chip, Nature 450, 272 (2007).

M. Asjad, F. Saif, Steady-state entanglement of a Bose-Einstein condensate and a nanomechanical resonator, Phys. Rev. A 84, 033606 (2011).

D. Teufel, T. Donner, D. Li, J. W. Harlow, M. S. Allman, K. Cicak, *et al.* Sideband cooling of micromechanical motion to the quantum ground state, Nature 475, 359 (2011).

F. Brennecke, S. Ritter, T. Donner, T. Esslinger, Cavity optomechanics with a Bose-Einstein condensate, Science 322, 235 (2008).

G. Szirmai, D. Nagy, and P. Domokos, Quantum noise of a Bose-Einstein condensate in an optical cavity, correlations, and entanglement, Phys. Rev. A 81, 043639 (2010).

A. B. Bhattacherjee, Quantum noise reduction using a cavity with a Bose-Einstein condensate, J. Phys. B 43, 205301 (2010).

S. K. Steinke, P. Meystre, Role of quantum fluctuations in the optomechanical properties of a Bose-Einstein condensate in a ring cavity, Phys. Rev. A 84, 023834 (2011).

H. Jing, Y. Jiang, W. P. Zhang, and P. Meystre, Laser-catalyzed spin-exchange process in a Bose-Einstein condensate, Phys. Rev. A 81, 031603(R) (2010).

X. D. Zhao, Z. W. Xie, and W. P Zhang, Modulational instability of nonlinear spin waves in an atomic chain of spinor Bose-Einstein condensates, Phys. Rev. B 76, 214408 (2007).

B. Chen, C. Jiang, and K. D. Zhu, Slow light in a cavity optomechanical system with a Bose-Einstein condensate, Phys. Rev. A 83, 055803 (2011a).

B. Chen, C. Jiang, J. J. Li and K. D. Zhu, All-optical transistor based on a cavity optomechanical system with a Bose-Einstein condensate, Phys. Rev. A 84, 055802 (2011b).

B. Chen, J. J. Li, C. Jiang and K. D. Zhu, Single photon router in the optical regime based on a cavity optomechanical system, IEEE Photonic Tech. Lett. 24, 766 (2012c).

B. Chen, C. Jiang, and K. D. Zhu, Tunable all-optical Kerr switch based on a cavity optomechanical system with a Bose-Einstein condensate, J. Opt. Soc. Am. B 28, 2007 (2011d).

K. W. Murch, K. L. Moore, S. Gupta, and D. M. Stamper-Kurn, Observation of quantum-measurement backaction with an ultracold atomic gas, Nature Phys. 4, 561 (2008).

M. Paternostro, G. De Chiara, and G.-M. Palma, Cold-atom-induced control of an optomechanical device, Phys. Rev. Lett. 104, 243602 (2010).

C. W. Gardiner and P. Zoller, Quantum Noise, Springer (2004).

S. Weis, R. Rivière, S. Deléglise, E. Gavartin, O. Arcizet, A. Schliesser, T. J. Kippenberg, Optomechanically induced transparency, Science 330, 1520 (2010).

A. H. Safavi-Naeini, T. P. Mayer Alegre, J. Chan, M. Eichenfield, M. Winger, Q. Lin, J. T. Hill, D. E. Chang, and O. Painter, Electromagnetically induced transparency and slow light with optomechanics, Nature 472, 69-73 (2011).

J. D. Teufel, D. Li, M. S. Allman, K. Cicak, A. J. Sirois, J. D. Whittaker, and R. W. Simmonds, Circuit cavity electromechanics in the strong-coupling regime, Nature 471, 204 (2011).

D. A. B. Miller, Are optical transistors the logical next step? Nat. Photonics 4, 3 (2010).

M. Soljačić and J. D. Joannopoulos, Enhancement of nonlinear effects using photonic crystals, Nature 3, 211 (2004).

M. Orrit, Photons pushed together, Nature 460, 42 (2009).

H. M. Gibbs, Optical Bistability: Controlling Light with Light (Academic, Orlando, 1985).

J. L. O'Brien, Optical quantum computing, Science 318, 1567 (2007).

D. Bouwmeester, A. Ekert, and A. Zeilinger, The physics of quantum information (Springer, Berlin, 2000).

D. E. Chang, A. S. Sørensen, E. A. Demler, and M. D. Lukin, A single-photon transistor using nanoscale surface plasmons, Nature Phys. 3, 807 (2007).

J. Hwang, M. Pototschnig, R. Lettow, G. Zumofen, A. Renn, S. Götzinger and V. Sandoghdar, A single-molecule optical transistor, Nature 460, 76 (2009).

G. S. Agarwal and S. Huang, Electromagnetically induced transparency in mechanical effects of light, Phys. Rev. A 81, 041803 (2010).

V. Fiore, Y. Yang, M. C. Kuzyk, R. Barbour, L. Tian, H. Wang, Storing optical information as a mechanical excitation in a silica optomechanical resonator, Phys. Rev. Lett. 107, 133601 (2011).

H. J. Kimble, The quantum internet, Nature, 453, 1203-1230 (2008).

T. Aoki, A. S. Parkins, D. J. Alton, C. A. Regal, B. Dayan, E. Ostby, K. J. Vahala, and H. J. Kimble, Efficient routing of single Pphotons by one atom and a microtoroidal cavity, Phys. Rev. Lett. 102, 083601 (2009).

I. C. Hoi, C. M. Wilson, G. Johansson, T. Palomaki, B. Peropadre, and P. Delsing, Demonstration of a single-photon router in the microwave regime, Phys. Rev. Lett. 107, 073601 (2011).

G. S. Agarwal and S. Huang, Optomechanical systems as single photon routers, Phys. Rev. A 85, 021801(R) (2012).

Y. J. Wang, D. Z. Anderson, V. M. Bright, E. A. Cornell, Q. Diot, T. Kishimoto, M. Prentiss, R. A. Saravanan, S. R. Segal, and S. Wu, Atom Michelson Interferometer on a chip using a Bose-Einstein condensate, Phys. Rev. Lett. vol. 94, 090405 (2005).

J. Fortágh and C. Zimmermann, Magnetic microtraps for ultracold atoms, Rev. Mod. Phys. 79, 235-289 (2007).

P. Treutlein, D. Hunger, S. Camerer, T. W. Hänsch, and J. Reichel, Bose-Einstein condensate coupled to a nanomechanical resonator on an atom chip, Phys. Rev. Lett. 99, 140403 (2007).

A. M. C. Dawes, L. Illing, S. M. Clark, and D. J. Gauthier, All-optical switching in rubidium vapor, Science, 308, 672-674 (2005).

R. W. Keyes, Power dissipation in information processing, Science 168, 796-801 (1970).

H. Schmidt and A. Imamoğlu, Giant Kerr nonlinearities obtained by electromagnetically induced transparency, Opt. Lett. 21, 1936-1938 (1996).

T. Tanabe, M. Notomi, S. Mitsugi, A. Shinya, and E. Kuramochi, All-optical switches on a silicon chip realized using photonic crystal nanocavities, Appl. Phys. Lett. 87, 151112 (2005).

J. Zhang, G. Hernandez, and Y. Zhu, All-optical switching at ultralow light levels, Opt. Lett. 32, 1317-1319 (2007).

M. Bajcsy, S. Hofferberth, V. Balic, T. Peyronel, M. Hafezi, A. S. Zibrov, V. Vuletic, and M. D. Lukin, Efficient all-optical switching using slow light within a hollow fiber, Phys. Rev. Lett. 102, 203902 (2009).

J. Scheuer, A. A. Sukhorukov, and Y. S. Kivshar, All-optical switching of dark states in nonlinear coupled microring resonators, Opt. Lett. 35, 3712-3714 (2010).

F. Qin, Y. Liu, Z. M. Meng, and Z. Y. Li, Design of Kerr-effect sensitive microcavity in nonlinear photonic crystal slabs for all-optical switching, J. Appl. Phys. 108, 053108 (2010).

X. Wei, J. Zhang, and Y. Zhu, All-optical switching in a coupled cavity-atom system, Phys. Rev. A. 82, 033808 (2010).

D. Nagy, P. Domokos, A. Vukics, and H. Ritsch, Nonlinear quantum dynamics of two BEC modes dispersively coupled by an optical cavity, Eur. Phys. J. D 55, 659-668 (2009).

C. W. Gardiner and M. J. Collett, Input and output in damped quantum systems: Quantum stochastic differential equations and the master equation, Phys. Rev. A 31, 3716-3774 (1985).

R. W. Boyd, Nonlinear Optics, (Academic Press, Amsterdam) p. 207 (2008).

A. B. Bhattacherjee, Cavity quantum optomechanics of ultracold atoms in an optical lattice: Normal-mode splitting, Phys. Rev. A 80, 043607 (2009).

H. J. Kimble, Strong interactions of single atoms and photons in cavity QED, Phys. Scripta. T76, 127-137 (1998).

J. D. Teufel, Dale Li, M. S. Allman, K. Cicak, A. J. Sirois, J. D. Whittaker, and R. W. Simmonds, Circuit cavity electromechanics in the strong-coupling regime, Nature 471, 204-208 (2011).

Chapter 5

The Smallest Generalized Optomechanical System — a Single Quantum Dot

Due to the unique size-dependent, narrow, symmetric, bright, and stable fluorescence, the nanoscale semiconductor quantum dots (SQD) offer unique opportunity to investigate sophisticated quantum optical effects in solid-state systems [Godden (2012); Moelbjerg (2012); Matthiesen (2012)]. Behaving as a simple stationary atom with good interference property, SQD lays the foundation for numerous possible applications, such as quantum interference [Bonadeo (1998)], Rabi oscillations [Patton (2005); Htoon (2002); Stievater (2001)], quantum computing [Yuan (2002)] and quantum repeaters [Briegel (1998)].

It has been demonstrated theoretically [Hughes (2012); Vlack (2012)] and shown experimentally [Moelbjerg (2012); Lermer (2012)] that in an ideal two-level or three-level SQD system, the strong coupling between light and matter leads to interesting spectral features, including absorption, ac-Stark effect and gain in the absorption spectrum driven by optical fields. For example, Muller and coauthors [Muller (2007)] have observed the resonance fluorescence from a coherently driven semiconductor quantum dot in a cavity. They treated the dynamics of a single quantum dot together with an external, nearly resonant electric field as two-level optical Bloch equations. Recently, Xu *et al.* [Xu (2007, 2008)] have experimentally observed Autler-Townes splitting and complex Mollow-related phenomenon in a single quantum dot system, using pump-signal technique.

In this chapter, we shall demonstrate that a single quantum dot can act as the smallest optomechanical system, while taking into account the role of lattice vibrations. In nanocavity optomechanical system with a single quantum dot, the exciton substitutes the optical cavity, while the mechanical oscillator is replaced by the lattice vibrations. At first, we theoretically simulate the experimental works done by Xu *et al.* [Xu (2007, 2008)] with their realistic parameters while neglecting exciton-phonon interaction. Since getting the same curves with Xu *et al.*, we next consider the impact of lattice vibrations of quantum dot and predict some new features in the signal absorption spectrum, due to phonon induced coherent population oscillation. Either the vibrational frequency or the lifetime of phonons plays an important role in the single quantum dot GOS. Here, we also show that, under certain conditions, the two steep peaks occurring in the signal absorption spectrum provide us an all-optical method to measure the vibrational frequency of LO-phonon.

At the last part of this chapter, we demonstrate the "optomechanically induced transparency" or phonon induced transparency (PIT) in a single quantum dot system, under the radiation of two optical fields. Together with the electromagnetically induced absorption (EIA) and parametric amplification (PA), we illustrate that a single quantum dot system can act as a quantum optomechanical transistor, which could cost less and be fabricated easily, and allow the realization of more sensitive and smaller optical repeater in quantum information technology. As we know, a single quantum dot belongs to the condensed matter, which means the generalized optomechanical system not only targets at optical materials, but also is concerned with solid state system. The smallest GOS — a single quantum dot proposed here means that the optomechanical system will soon evolve into condensed matter optomechanics and help to significantly improve the miniaturization and integration.

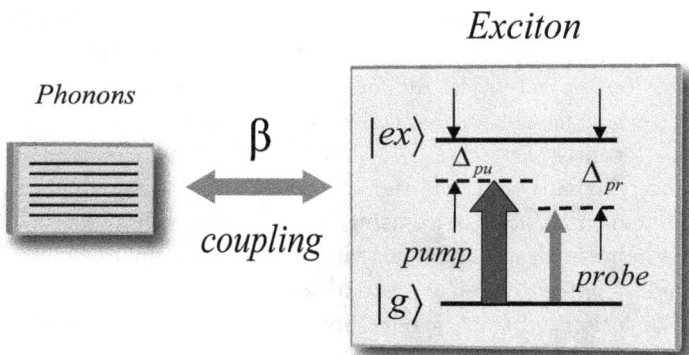

Fig. 5.1   Energy levels for a single quantum dot system, where the two-level exciton is coupled with the vibrational modes of LO-phonon.

The atomic two-level system has already been investigated by Yu *et al.* [Yu (1997)], where the emission spectrum of Ba atoms were investigated driven by continuous bichromatic field. In the previous works [Xu (2007, 2008); Zrenner (2002); Stufler (2005)], the semiconductor quantum dot can be modeled as a two-level system, which consists of the ground state (no exciton) $|g\rangle$ and the first excited state (single exciton) $|ex\rangle$ (as shown in Fig.5.1). However, the two-level energy levels will have a splitting while the exciton dressing with the LO-phonons. Here, only LO-phonons with small momenta can efficiently couple to the exciton [Inoshita (1992)]. As usual, this two-level system can be charaterized by the pseudospin -1 operators $\sigma^{\pm}$ and σ^z. Therefore, the Hamiltonian of this two-level exciton is given by

$$H_{\text{ex}} = \hbar\omega_{\text{ex}}\sigma^z, \tag{5.1}$$

where ω_{ex} is the frequency of exciton.

The lattice vibrations can be described by the Hamiltonian

$$H_{\text{ph}} = \sum_{\vec{k}} \hbar\omega_{\text{ph}} f_{\vec{k}}^+ f_{\vec{k}}, \tag{5.2}$$

where ω_{ph} is the vibrational frequency of phonons, $f_{\vec{k}}^+$ and $f_{\vec{k}}$ are the creation and annihilation operators ($\hbar\vec{k}$ corresponds to the phonon momentum), respectively.

In the simultaneous presence of a strong pump field and a weak signal field, the Hamiltonian for a single quantum dot system can be written as [Wilson-Rae (2002); Boyd (2008)]

$$H = H_{\text{ex}} + H_{\text{ph}} + H_{\text{ex-ph}} + H_{\text{ex-o}}, \tag{5.3}$$

$$H_{\text{ex-ph}} = \hbar\sigma^z \sum_{\vec{k}} \lambda_{\vec{k}}(f_{\vec{k}}^+ + f_{\vec{k}}), \tag{5.4}$$

$$\begin{aligned} H_{\text{ex-o}} = &-\mu(\sigma^+ E_p e^{-i\omega_p t} + \sigma^- E_p^* e^{i\omega_p t}) \\ &- \mu(\sigma^+ E_s e^{-i\omega_s t} + \sigma^- E_s^* e^{i\omega_s t}), \end{aligned} \tag{5.5}$$

where $H_{\text{ex-ph}}$ represents the interaction between exciton and phonons, $\lambda_{\vec{k}}$ represents the coupling strength. In this system. $H_{\text{ex-o}}$ describes the exciton coupling to the two optical fields, E_p and E_s are slowly varying envelope of the strong pump field and weak signal field, respectively. μ is the electric dipole moment of the exciton. In a frame rotating at the pump field frequency ω_p, the total Hamiltonian reads as follows.

$$H = \hbar\Delta_p\sigma^z + \hbar\omega_{\mathrm{ph}}\sum_{\vec{k}}f_{\vec{k}}^+ f_{\vec{k}} + \hbar\sigma^z\sum_{\vec{k}}\lambda_{\vec{k}}(f_{\vec{k}}^+ + f_{\vec{k}})$$
$$-\hbar\Omega(\sigma^+ + \sigma^-) - \mu(\sigma^+ E_s e^{-i\delta t} + \sigma^- E_s^* e^{i\delta t}), \qquad (5.6)$$

where $\Delta_p = \omega_{\mathrm{ex}} - \omega_p$ and $\delta = \omega_s - \omega_p$ represent the pump-exciton detuning and the signal-pump detuning, respectively. $\Omega = \mu E_p/\hbar$ is the Rabi frequency of the pump field.

Here we use the quantum Heisenberg-Langevin approach to solve the Hamiltonian equation, which can be written as [Zoller (2004); Scully (1997)]

$$\frac{d\sigma^z}{dt} = -\Gamma_1(\sigma^z + 1) + i\Omega(\sigma^+ - \sigma^-) + \frac{i\mu}{\hbar}(E_s e^{-i\delta t}\sigma^+ - E_s^* e^{i\delta t}\sigma^-), \quad (5.7)$$

$$\frac{d\sigma^-}{dt} = -(\Gamma_2 + i\Delta_p)\sigma^- - iq\sigma^- - 2i\Omega\sigma^z - \frac{2i\mu E_s e^{-i\delta t}}{\hbar}\sigma^z + \hat{F}_n, \qquad (5.8)$$

$$\frac{d^2}{dt^2}q + \gamma_{\mathrm{ph}}\frac{d}{dt}q + \omega_{\mathrm{ph}}^2 q = -2\omega_{\mathrm{ph}}^3\beta\sigma^z, \qquad (5.9)$$

where Γ_1 and Γ_2 are the exciton spontaneous emission rate and dephasing rate, respectively; $\beta = \sum_{\vec{k}}\lambda_{\vec{k}}^2/\omega_{\mathrm{ph}}^2$ is the Huang-Rhys factor which corresponds to the exciton-phonons coupling [Huang (1950)]; $q = \Sigma_{\vec{k}}\lambda_{\vec{k}}(f_{\vec{k}}^+ + f_{\vec{k}})$ is the position operator of the phonons; γ_{ph} is the phonon decay rate and $\hat{F}_n$ is the δ-correlated Langevin noise operator which has zero mean $\langle\hat{F}_n\rangle = 0$ and correlation $\langle\hat{F}_n^+(t)\hat{F}_n(t')\rangle \sim \delta(t - t')$.

In the presence of a strong pump field, we consider the mean response of a single quantum dot to the signal field, neglecting the quantum fluctuations [Weis (2010)]. This is similar to what has been treated in electromagnetically induced transparency (EIT) where the atomic mean value equations are used and all the quantum fluctuations are neglected, due to both spontaneous emission and collisions [Hau (1999)]. By factorizing the phonons and exciton degrees of freedom, i.e., $\langle q\sigma^-\rangle = \langle q\rangle\langle\sigma^-\rangle$, in which any entanglement between them should be neglected, and then Eqs.(5.7)-(5.9) are given by

$$\frac{d\langle\sigma^z\rangle}{dt} = -\Gamma_1(\langle\sigma^z\rangle + 1) + i\Omega(\langle\sigma^+\rangle - \langle\sigma^-\rangle) + \frac{i\mu}{\hbar}(E_s e^{-i\delta t}\langle\sigma^+\rangle$$
$$- E_s^* e^{i\delta t}\langle\sigma^-\rangle), \qquad (5.10)$$

$$\frac{d\langle\sigma^-\rangle}{dt} = -(\Gamma_2 + i\Delta_p)\langle\sigma^-\rangle - i\langle q\rangle\langle\sigma^-\rangle - 2i\Omega\langle\sigma^z\rangle - \frac{2i\mu E_s e^{-i\delta t}}{\hbar}\langle\sigma^z\rangle,$$

(5.11)

$$\frac{d^2}{dt^2}\langle q\rangle + \gamma_{\text{ph}}\frac{d}{dt}\langle q\rangle + \omega_{\text{ph}}^2\langle q\rangle = -2\omega_{\text{ph}}^3\beta\langle\sigma^z\rangle.$$

(5.12)

In order to solve Eqs.(5.10)-(5.12), as usual, we make the ansatz [Boyd (2008)]: $\langle\sigma^-(t)\rangle = \sigma_0 + \sigma_+ e^{-i\delta t} + \sigma_- e^{i\delta t}$, $\langle\sigma^z(t)\rangle = \sigma_0^z + \sigma_+^z e^{-i\delta t} + \sigma_-^z e^{i\delta t}$, $\langle q(t)\rangle = q_0 + q_+ e^{-i\delta t} + q_- e^{i\delta t}$. By substituting them into Eqs.(5.10)-(5.12) and working to the lowest order in E_s, but to all orders in E_p, we finally obtain the linear optical susceptibility in the steady state as the following solution

$$\chi_{\text{eff}}^{(1)}(\omega_s) = \frac{\mu\sigma_+}{E_s} = \frac{\mu^2}{\hbar\Gamma_2}\chi^{(1)}(\omega_s),$$

(5.13)

where the dimensionless susceptibility is given by

$$\chi^{(1)}(\omega_s) = \frac{w_0}{(-\Delta_{\text{p0}} + \omega_{\text{ph0}}\beta w_0 + i + \delta_0)} +$$

$$\frac{2w_0\Omega_p^2(2i + \delta_0)[-\Delta_{\text{p0}} + i - \omega_{\text{ph0}}\beta w_0(\xi(\omega_s) + 1)][(\Delta_{\text{p0}} - \omega_{\text{ph0}}\beta w_0 + i + \delta_0)]}{\phi(\delta_0)}.$$

(5.14)

The auxiliary function $\xi(\omega_s)$ and the function $\phi(\delta_0)$ are given by

$$\xi(\omega_s) = \frac{\omega_{\text{ph0}}^2}{\omega_{\text{ph0}}^2 - i\delta_0\gamma_{\text{ph0}} - \delta_0^2},$$

(5.15)

$$\phi(\delta_0) = (\Delta_{\text{p0}} - \omega_{\text{ph0}}\beta w_0 - i - \delta_0)\times$$
$$\{[(i + \delta_0)^3 - (i + \delta_0)(\Delta_{\text{p0}} - \omega_{\text{ph0}}\beta w_0)^2][(\Delta_{\text{p0}} - \omega_{\text{ph0}}\beta w_0)^2 + 1]$$
$$+ 2\Omega_p^2[\Delta_{\text{p0}} - \omega_{\text{ph0}}\beta w_0 + i + \omega_{\text{ph0}}\beta\xi(\omega_s)w_0]$$
$$[1 - i\delta_0 - (\Delta_{\text{p0}} - \omega_{\text{ph0}}\beta w_0)(2i + \delta_0)]\},$$

(5.16)

where $\gamma_{\text{ph0}} = \gamma_{\text{ph}}/\Gamma_2$, $\delta_0 = \delta/\Gamma_2$, $w_0 = \sigma_0^z$, $\Gamma_1 = 2\Gamma_2$, $\omega_{\text{ph}} = \omega_{\text{ph}}/\Gamma_2$, $\Omega_p = \Omega/\Gamma_2$, $\Delta_{\text{p0}} = \Delta_p/\Gamma_2$.

The population inversion w_0 of the exciton is determined by the following equation

$$(w_0 + 1)[(\Delta_{\text{p0}} - \beta\omega_{\text{ph0}}w_0)^2 + 1] + 2\Omega_p^2 w_0 = 0.$$

(5.17)

5.1 Two hallmarks of a single quantum dot as generalized optomechanical system

The form of cubic Eq.(5.17) is characteristic of the optical multistability [Larson (2008); Gupta (2007)]. Figure 5.2 shows the steady-state value of the population inversion w_0 as a function of the pump-exciton detuning (a) for fixed values of pump Rabi frequency, and (b) as a function of the pump Rabi frequency for a fixed value of the pump-exciton detuning. As discussed in Chapter 1, the bistable steady state of Fig.5.2 shows that a single quantum dot can act as a generalized optomechanical system. The dashed curve and solid curve correspond to the unstable and stable behavior, respectively. In Fig.5.2(a), Here, we notice that the value of population inversion w_0 is confined between 0 and -1, which is totally different from the conventional cavity optomechanical system. In cavity optomechanical system, the input pump field directly affects the coupling between the cavity photons and the mechanical oscillator, where the photon number of the inside cavity is decided by the incident pump laser. However, for the single quantum dot GOS, the deformation coupling of exciton-phonon is independent on the pump fields. The population inversion of quantum dot can only take the value between 0 and -1 via the interaction between the exciton and the pump field.

In the presence of the strong pump laser, the OMIT window of the reflected signal laser with cavity optomechanical system arises from the

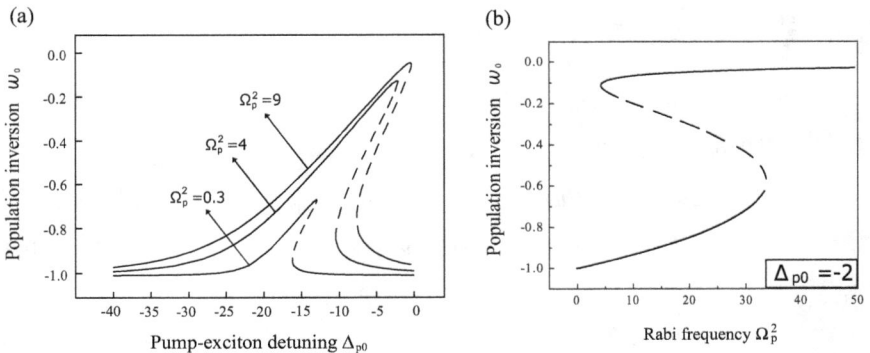

Fig. 5.2 (a) Steady-state population inversion w_0 as a function of pump-exciton detuning with different Rabi frequencies of the pump field. The dashed curves and solid curves correspond to the unstable states and table states, respectively. (b) Population inversion w_0 as a function of Rabi frequency of the pump laser. The parameters used are $\beta = 0.2$, $\omega_{\mathrm{ph0}} = 100$.

interaction between the photons and the mechanical resonator via radiation pressure [Weis (2010)]. However, in a single quantum dot based optomechanical system, the excitonic resonance couples to phonons via deformation potential that matches the cavity mode interacts with mechanical resonator via radiation pressure, which also demonstrates "optomechanically induced transparency" or called phonon induced transparency (PIT). Figure 5.3 shows the comparison between the conventional cavity optomechanical system and a single quantum dot based optomechanical system,

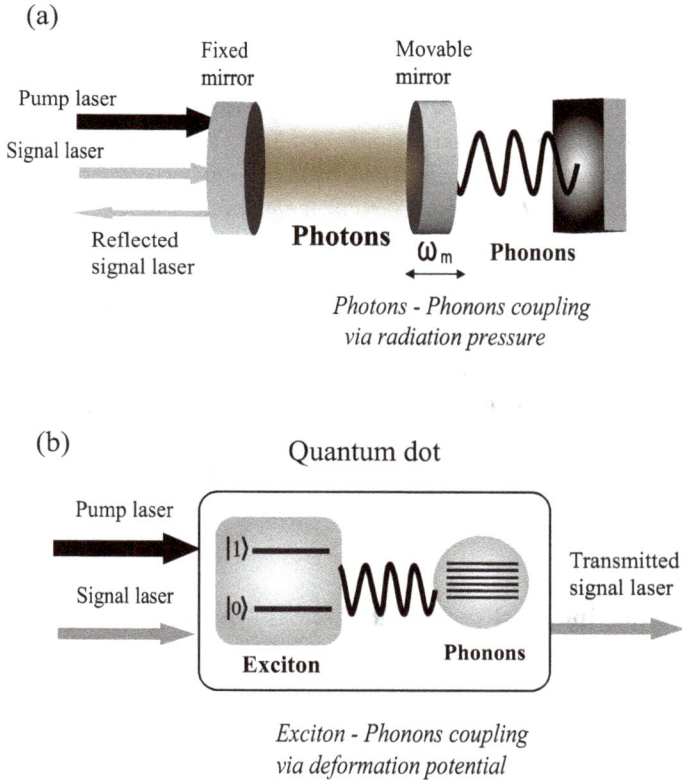

Fig. 5.3   Comparison of a cavity optomechanical system and a single quantum dot GOS. (a) Typical cavity optomechanical system. The left mirror is fixed, while the right movable mirror couples to the optical cavity field via radiation pressure. (b) A single quantum dot optomechanical system. The two-level exciton interacts with phonon modes via deformation potential in the presence of a strong pump laser and a weak signal laser. In contrast to the optomechanical system presented in (a), the light detection is the transmitted signal laser other than the reflected signal laser.

where the photon-phonon coupling in cavity optomechanical system is in analogy with the exciton-phonon interaction in a single quantum dot.

Figure 5.4 demonstrates the second hallmark of generalized optomechanical system, that is, the switch from optomechanically induced transparency, electromagnetically induced absorption (EIA) to parametric amplification. For illustration of the experimentally achievable parameters, we choose the realistic self-organized InAs/GaAs quantum dot. At room temperature, the dephasing time of exciton is 0.3ps [Sauvage (2002)], which corresponds to $\Gamma_2 = 3.3$THz, the lifetime and the frequency of LO-phonon are 12ps (i.e., $\gamma_{ph} = 83$GHz) and $\omega_{ph} = 50$THz [Zibik (2004)], respectively, the Huang-Rhys factor β is 0.015 [Heitz (1999)].

Figure 5.4(a) shows the "optomechanically induced transparency" or phonon induced transparency in a single quantum dot system where the exciton is driven on its red sideband, i.e., $\Delta_p = \omega_{ph}$, which is very similar to the experimental results obtained by Kippenberg *et al.* [Weis (2010)]. However, if switching the pump-exciton detuning on the blue sideband, i.e., $\Delta_p = -\omega_{ph}$, the signal transmission curve become a dip, as shown in Fig.5.4(b). The greater the intensity of pump filed, the deeper the peak value. Figure 5.4(b) displays this situation, where the negative transmission of the laser signal is increased with the increase of the pump power. This is the so-called electromagnetically induced absorption effect [Lezama (1999)]. However, as the pump field intensity increases even further, the system switches from EIA to PA resulting in the signal field amplification (see Fig.5.4(c)), which is the same with that in the cavity optomechanical system [Safavi-Naeini (2011)].

5.2   Phonon induced coherent optical spectroscopy

In the following, we will present numerical results of the signal absorption spectrum. For the realistic quantum dots such as self-organized InAs/GaAs, the Huang-Rhys factor β is 0.02 [Heitz (1999)], the LO-phonon energy is 36meV which corresponds to LO-phonon frequency $\omega_{ph} = 37800$GHz [Mahan (1981)]. The dephasing time of exciton is $T_2 = 0.3$ps ($T_2 = 1/\Gamma_2$) and the of LO-phonon is $\tau_{ph} = 12$ps ($\tau_{ph} = 1/\gamma_{ph}$), respectively [Sauvage (2002); Zibik (2004)].

Figure 5.5 shows the signal absorption spectrum without and with considering the role of LO-phonons, while fixing the pump field on the resonance of exciton ($\Delta_p = 0$). In the absence of coupling between exciton and

(a)

(b)

(c)

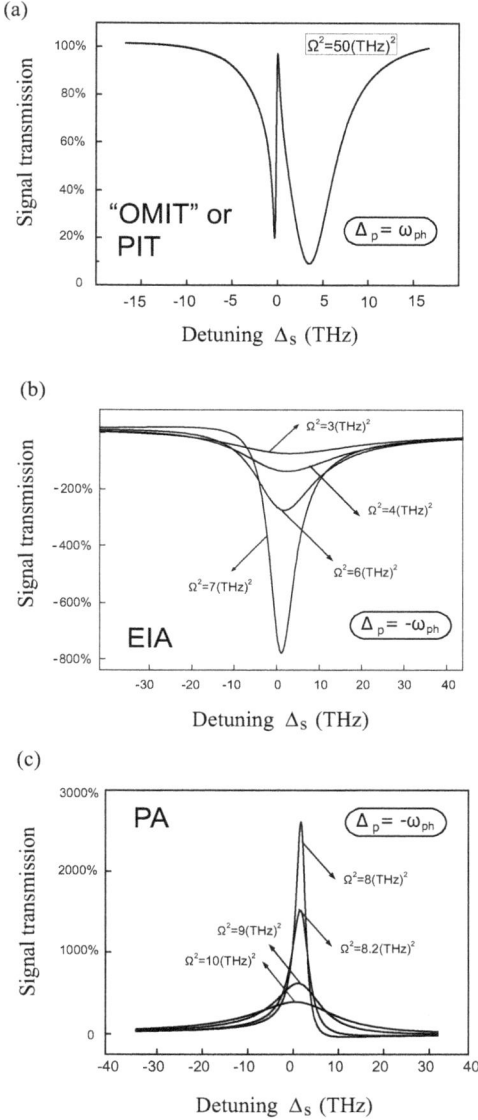

Fig. 5.4   The second hallmark of GOS. Plots of optomechanically induced transparency (OMIT) or phonon induced transparency (PIT), electromagnetically induced absorption (EIA) and parametric amplification (PA). Signal field transmission as a function of signal-exciton detuning for (a) $\Delta_p = \omega_{\mathrm{ph}}$; (b) $\Delta_p = -\omega_{\mathrm{ph}}$ with small Rabi frequencies of the pump field; (c) $\Delta_p = -\omega_{\mathrm{ph}}$ with large Rabi frequencies of the pump field. The realistic parameters used are $\beta = 0.015$, $\gamma_{\mathrm{ph}} = 83\mathrm{GHz}$ and $\Gamma_2 = 3.3\mathrm{THz}$.

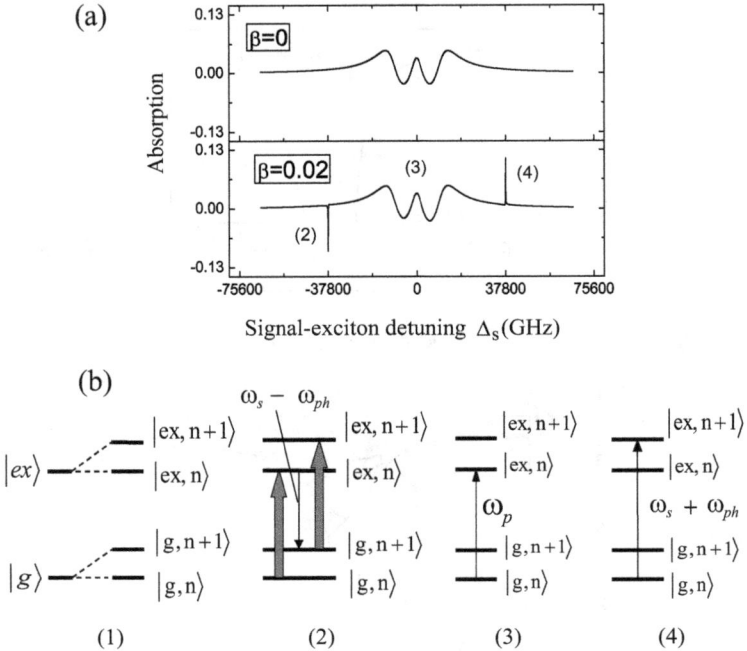

Fig. 5.5   (a) The absorption spectrum of a signal field without and with the coupling between exciton and phonons for the case $\Omega_R^2 = 25(\text{THz})^2$, $\Delta_p = 0$, $\omega_{\text{ph}} = 37800\text{GHz}$ and $\tau_{\text{ph}} = 12\text{ps}$. (b) The new features in the signal absorption spectrum shown in (a) are identified by the corresponding transitions between the dressed states, where $|n\rangle$ denotes the number states of phonon modes.

LO-phonons ($\beta = 0$), the top curve in Fig.5.5(a) has a excellent agreement with Xu *et al.*'s experiment [Xu (2007)], where only a Mollow absorption effect in signal absorption spectrum. However, at the coupling on situation ($\beta = 0.02$), the bottom of Fig.5.5(a) gives a novel absorption spectrum. Here the coherent lattice vibration brings some new features that are different from those in atomic-like systems without phonons. It is shown that two sharp peaks appearing at the points $\pm\omega_{\text{ph}}$ are related to the phonon-induced coherent population oscillation, which are different from those obtained by Xu *et al.* [Xu (2007)]. Figure 5.5(b) shows energy levels and transitions between exciton and phonon mode while applying two optical fields. Part (1) shows that the original two levels of exciton split into four dressed state while coupling with the phonon modes. Part (2) represents the phonon induced three-photon resonance, where the exciton makes a transition from the lowest dressed level to the highest dressed level by the simultaneous absorption of two photons and the emission of a photon at $\omega_p - \omega_{ph}$. This process can amplify a wave at $\Delta_s = -\omega_{ph} = -37800\text{GHz}$, which is indicated by the region of negative absorption of the bottom plot in Fig.5.5(a). Part (3) shows the origin of phonon induced stimulated Rayleigh resonance, which corresponds to a transition from the lowest dressed level $|g, n\rangle$ to the dressed level $|ex, n\rangle$. The rightmost part (4) is related to the phonon induced absorption resonance as modified by the ac-Stark effect, which corresponds to $\Delta_s = \omega_{ph} = 37800\text{GHz}$.

We further investigate the signal absorption spectrum in a single quantum dot system, while the pump field is detuned from the exciton frequency ($\Delta_p \neq 0$). Figure 5.6 shows the signal absorption spectrum with various pump detuning Δ_p as a function of signal-exciton detuning Δ_s without (Fig.5.6(a)) and with (Fig.5.6(b)) the coupling between exciton and LO-phonons. On one hand, if we set the coupling strength β to zero (Fig.5.6(a)) and scan the signal field across the exciton frequency ω_{ex}, only the familiar Mollow curves turned up in the signal absorption spectrum, which matches the curves in [Xu (2008)]. On the other hand, when the coupling turns on, two steep peaks appeared in the signal absorption spectrum, due to phonon induced coherent population oscillation, as shown in Fig.5.6(b). It should be noticed that these two peaks are shifted compared with the bottom curve in Fig.5.5(a). The offset is determined by the detuning Δ_p, which means, the two steep sidebands are located at the pump-signal detuning $\delta = \Delta_s + \Delta_p = \pm\omega_{ph}$. The inset of Fig.5.6(b) is the enlarged view of the negative peak, which satisfies $\Delta_s + \Delta_p = -\omega_{ph}$. The solid curve in Fig.5.6 is the same as the bottom plot in Fig.5.5(a), i.e., $\Delta_p = 0$, which locates at $\Delta_s = -\omega_{ph} = -37800\text{GHz}$ exactly.

From Fig.5.5 and Fig.5.6, we can conclude that either on the resonance $\Delta_p = 0$ or the detuned $\Delta_p \neq 0$ situation, the two peaks at the both sides will always be around $\pm\omega_{ph}$ in the signal absorption spectrum. But which factors will affect the intensity of the peak? Figure 5.7 gives us an answer.

The absorption spectrum as a function of signal-exciton detuning Δ_s with different phonon lifetimes ($\tau_{ph} = 1\text{ps}$, 5ps and 12ps) are plotted in Fig.5.7. Apparently, the longer the phonon lifetime is, the higher the peak will be. Here the lifetime (τ_{ph}) is inversely proportional to the decay rate of phonon (γ_{ph}). We noticed that when the lifetime of phonon reducing to 1ps, the two steep peaks at the both sides in the signal absorption spectrum are difficult to be recognized.

(a)   Coupling off

Signal-exciton detuning Δ_s (GHz)

Coupling on

(b)

Signal-exciton detuning Δ_s (GHz)

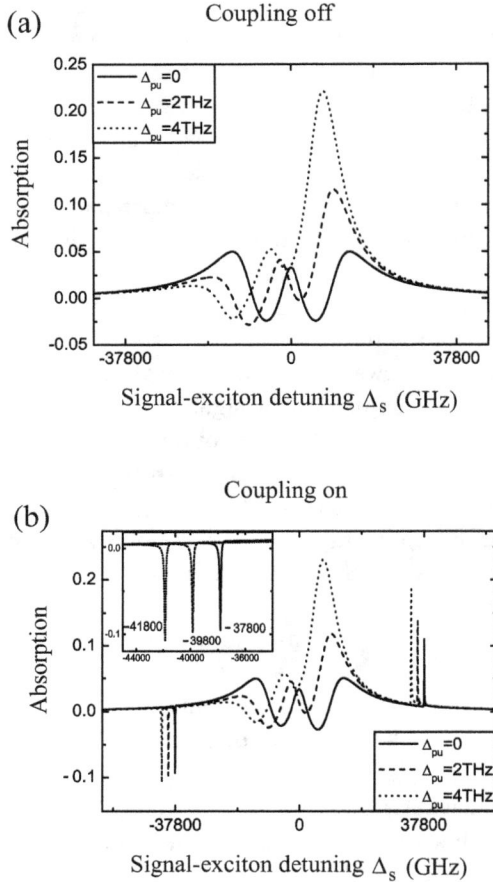

Fig. 5.6   The signal absorption spectrum as a function of the signal-exciton detuning Δ_s with different pump detuning Δ_p, while (a) neglecting and (b) considering the coupling between exciton and LO-phonon modes. The inset window of Fig.5.6(b) is the enlarged view of negative region around -37800GHz. The other parameters are $\omega_{\mathrm{ph}} = 37800$GHz and $\tau_{\mathrm{ph}} = 12$ps, $\Omega_R^2 = 25(\mathrm{THz})^2$.

5.3   Measurement the frequency of LO-phonon

If observing the curves in Fig.5.7 carefully, we can notice that the two steep peaks at the both sides of signal absorption spectrum just correspond to the vibrational frequency of LO-phonon. Which means, the resonance situation, i.e., $\Delta_p = 0$ provides a simple optical method to measure the

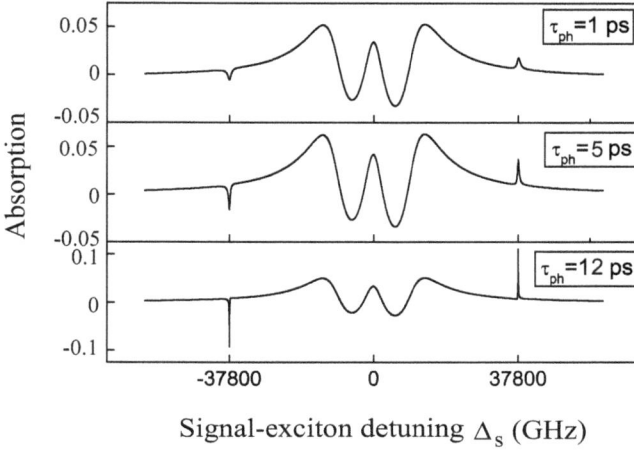

Fig. 5.7 The signal absorption spectrum with different lifetime of phonon. The other realistic parameters are $\omega_{\mathrm{ph}} = 37800\mathrm{GHz}$ and $\Delta_p = 0$, $\Omega_R^2 = 25(\mathrm{THz})^2$, $\beta = 0.02$.

frequency of LO-phonon. Therefore, if one fixes the frequency of pump field on the resonant of exciton and scan the second signal field across the exciton frequency ω_{ex}, then we can easily obtain the LO-phonon frequency at $\Delta_s = \pm\omega_{\mathrm{ph}}$ in the signal absorption spectrum. However, as discussed in Fig.5.7, the lifetime of LO-phonon plays an important role in the phonon induced coherent population oscillation. If the lifetime of phonon is too short (such as at room temperature), the two steep peaks at the both sides of signal absorption spectrum are very likely to be interfered by environment noise and eventually annihilated in experiment. This is the reason why many scientists have not observed phonon peaks in laboratory.

5.4 Slow light and fast light

In terms of a single quantum dot system, we determine the light group velocity as [Bennink (2001); Harris (1992)]

$$v_g = \frac{c}{n + \omega_s(dn/d\omega_s)}, \tag{5.18}$$

where $n \approx 1 + 2\pi\chi_{\mathrm{eff}}^{(1)}$, and then

$$\frac{c}{v_g} = 1 + 2\pi\mathrm{Re}\chi_{\mathrm{eff}}^{(1)}(\omega_s)_{\omega_s=\omega_{\mathrm{ex}}} + 2\pi\omega_s\mathrm{Re}(\frac{d\chi_{\mathrm{eff}}^{(1)}}{d\omega_s})_{\omega_s=\omega_{\mathrm{ex}}}. \tag{5.19}$$

It is clear from this expression for v_g that when $\mathrm{Re}\chi^{(1)}(\omega_s)_{\omega_s=\omega_{ex}}$ is zero and the dispersion is steeply positive or negative, the group velocity is significantly reduced or increased, and then the group velocity index n_g can be defined as

$$n_g = \frac{c}{v_g} - 1 = \frac{c - v_g}{v_g} = \frac{2\pi\omega_{ex}\rho\mu^2}{\hbar\Gamma_2}\mathrm{Re}\left(\frac{d\chi^{(1)}(\omega_s)}{d\omega_s}\right)_{\omega_s=\omega_{ex}} \quad (5.20)$$

$$= \Gamma_2\Sigma\mathrm{Re}\left(\frac{d\chi^{(1)}(\omega_s)}{d\omega_s}\right)_{\omega_s=\omega_{ex}},$$

where $\Sigma = 2\pi\omega_{ex}\rho\mu^2/\epsilon_0\hbar\Gamma_2^2$. One can observe the slow light when $n_g > 0$, and the superluminal light when $n_g < 0$ [Gauthier (2009)].

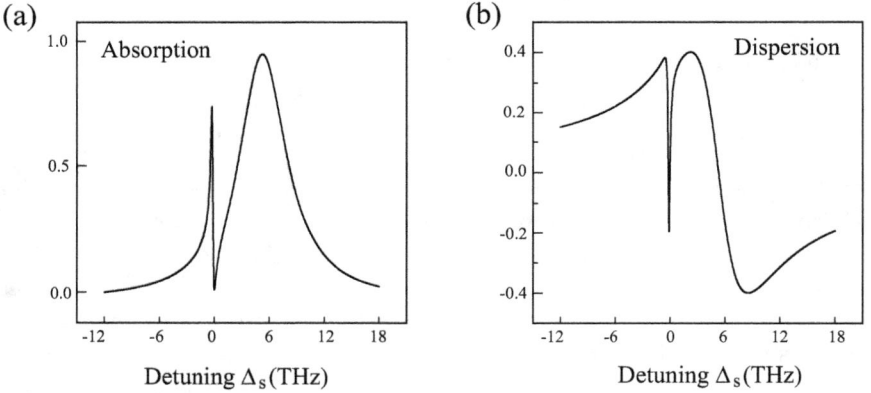

Fig. 5.8   The (a) imaginary part and (b) the real part of of $\chi^{(1)}(\omega_s)$, which correspond to the absorption and the dispersion of the signal field, respectively. The parameters used are $\Delta_p = \omega_{ph} = 50\mathrm{THz}$, $\Omega^2 = 10(\mathrm{THz})^2$, $\gamma_{ph} = 83\mathrm{GHz}$.

In the present of a strong pump filed, Fig.5.8 shows the imaginary part and the real part of of the signal field via passing through a single quantum dot system. In Chapter 3, we have demonstrated that the imaginary part and real part of the optical susceptibility $\chi^{(1)}(\omega_s)$ just correspond to the absorption and dispersion of the signal field, respectively. According to Eq.(5.20), the positive slope at $\Delta_s = 0$ (Fig.5.8(b)) means that, when $\Delta_p = \omega_{ph}$ there is a slow light effect in a single quantum dot, which accompanied with a zero absorption shown in Fig.5.8(a).

In Fig.5.9, we present numerical results of the group velocity index n_g for different Rabi frequencies, with $\Delta_p = \omega_{ph}$ and $\Delta_s = 0$. It is obvious that near $\Omega^2 = 5(\text{THz})^2$, the most slow light can be produced in a single quantum dot system. From the figure we can estimate the group velocity index for InAs/GaAs quantum dots as 25 for $T_2 = 300\text{fs}$. After $\Omega^2 = 5(\text{THz})^2$, the group velocity index n_g becomes small with increasing Ω^2. However, the magnitude of slow light is decided by the number density of quantum dot. Actually, in InAs/GaAs quantum dots, the base has a length of 25nm, a height of 2.5nm and an area density of about $4 \times 10^{10}\text{cm}^2$. The vertical stacking of the quantum dot layers consists of 40 InAs quantum dot layers separated by 35nm thick GaAs barriers. Then, the output single field can be 3×10^6 times slower than the input pulse.

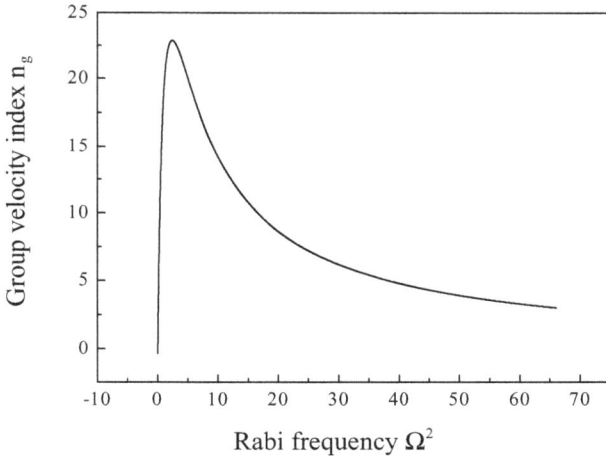

Fig. 5.9   The group velocity index of slow light (in units of Σ) as a function of the pump Rabi frequency Ω^2. The other parameters used are $\Delta_p = \omega_{\text{ph}} = 50\text{THz}$, $\gamma_{\text{ph}} = 83\text{GHz}$, $\Delta_s = 0$.

Equation (5.20) also displays that if the dispersion become a negative slope, there is a superliminal light, i.e., $n_g < 0$. Figure 5.10 plots the absorption curve and dispersion curve of the signal field with $\Delta_p = 0$, which denote the negative slope at $\Delta_s = 0$ in (b) and vanished absorption in (a). Then, we anticipate that there is a superluminal light in a single quantum dot system. Figure 5.11 plots the group velocity index n_g as a function of Rabi frequencies of the signal field with $\Delta_s = \Delta_p = 0$, which exhibits that for a single quantum dot GOS, the output signal field can be 9 times faster than the input pulse.

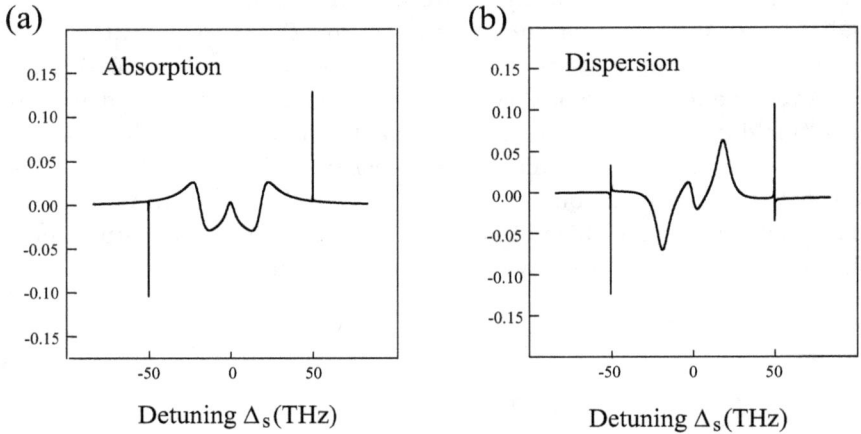

Fig. 5.10   The (a) absorption and (b) dispersion part of signal field with $\Delta_p = 0$. The other parameters used are $\Delta_p = 0$, $\omega_{ph} = 50$THz, $\Omega^2 = 10(\text{THz})^2$, $\gamma_{ph} = 83$GHz.

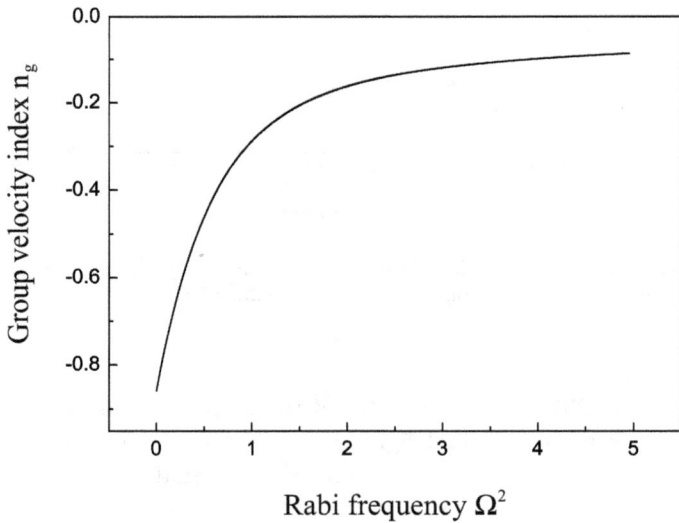

Fig. 5.11   The group velocity index of fast light (in units of Σ) as a function of the pump Rabi frequency Ω^2. The other parameters used are $\Delta_s = \Delta_p = 0$, $\omega_{ph} = 50$THz, $\gamma_{ph} = 83$GHz.

5.5 A quantum optical transistor

From Fig.5.4, we can see that the signal transmission can be modulated by the pump field, which provide a potential of an all-optical transistor. Here, we plot the relationship between signal transmission and pump laser intensity in Fig.5.12, which exhibits that there exists a turning point between $\Omega^2 = 7(\text{THz})^2$ and $\Omega^2 = 8(\text{THz})^2$. The turning point switches the signal transmission from EIA to PA. Turning on and off the pump field correspond to the attenuated and amplified single field, respectively. However, low case should be noticed that the maximum amplification goes to 8000% at $\Omega^2 = 7.5(\text{THz})^2$, which means the optomechanical system with a single quantum dot can not only switch the weak laser from off to on, but also serve as a laser amplifier via exciton-phonons interaction.

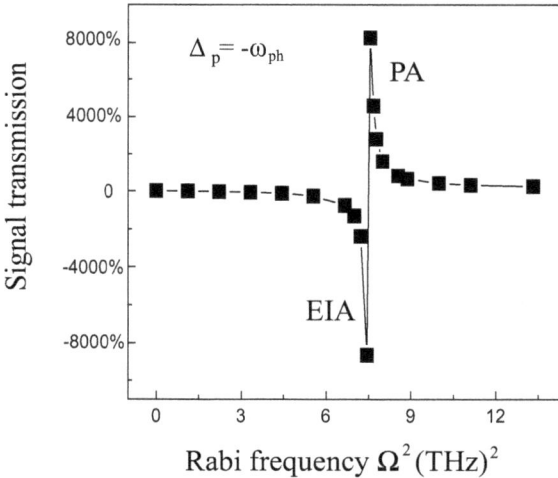

Fig. 5.12   The relationship between signal transmission and pump laser intensity. The parameters used are the same with Fig.5.4.

The gain of signal laser shown in Fig.5.12 with the region $\Omega^2 \geq 8 \ (\text{THz})^2$ gives us a special idea that the nanocavity optomechanics with a single quantum dot can act as a quantum optomechanical transistor. Figure 5.13 illustrates the physical protocol of this optomechanical transistor. Figure 5.13(b) displays the transmission spectrum when we shelve the pump field and only apply a signal field. This plot shows that in the absence of the pump field, a single quantum dot attenuates the weak signal field totally.

This dip arises from the usual exciton absorption resonance. However, as the pump field turns on and fixes the pump-exciton detuning $\Delta_p = -\omega_{\mathrm{ph}} = -50\mathrm{THz}$, the dip switches to a transmission peak immediately (see Fig.5.13(c)). Figure 5.13(c) displays that the transmission of the signal laser can reach at 600% when $\Omega^2 = 9(\mathrm{THz})^2$, which is generated by a single quantum dot. This amplification comes from the quantum interference between the phonons and the beat of the two optical fields via the exciton in the quantum dot, which has been discussed before. Since dressing with the phonon modes, the original two levels of exciton split into several metastable levels, then the electrons make transitions between them after applying a strong pump laser, which results in the constructive interference, and eventually amplifies the weak signal laser.

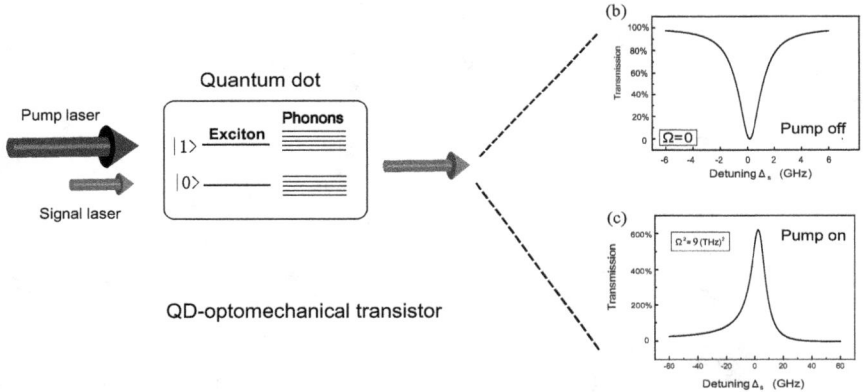

Fig. 5.13   Attenuation and amplification of the signal field when turning on and turning off the pump laser, respectively. The realistic parameters used are $\Delta_p = -\omega_{\mathrm{ph}} = -50\mathrm{THz}$, $\beta = 0.015$, $\gamma_{\mathrm{ph}} = 83\mathrm{GHz}$ and $\Gamma_2 = 3.3\mathrm{THz}$. (a) Schematic illustration of a single quantum dot based optomechanical transistor in the presence of a strong pump laser and a weak signal laser. (b) Attenuation of the signal laser while turning off the pump laser. (c) When turning on the pump laser, the attenuation dramatically becomes an amplification in a single quantum dot based optomechanical system.

Practically, as we know, there are many factors affecting the transmission spectrum of the quantum dot, such as the exciton dephasing, the decay rate of phonons and the exciton-phonon coupling. In order to achieve better experimental results, it is important to select the appropriate parameters of the quantum dot. Most recently, the other quantum dots based on graphene are fabricated in laboratory, which have the large exciton energy and the coupling between excitons and phonons seems stronger than that in the ordinary semiconductor quantum dot [Lu (2011)]. These graphene quantum dots may be the best choice to realize our scheme proposed here.

Bibliography

T. M. Godden, J. H. Quilter, A. J. Ramsay, Yanwen Wu, P. Brereton, S. J. Boyle, I. J. Luxmoore, J. Puebla-Nunez, A. M. Fox, and M. S. Skolnick, Coherent optical control of the spin of a single hole in an InAs/GaAs quantum dot, Phys. Rev. Lett. 108, 017402 (2012).

A. Moelbjerg, P. Kaer, M. Lorke, and J. Mørk, Resonance fluorescence from semiconductor quantum dots: beyond the Mollow triplet, Phys. Rev. Lett. 108, 017401 (2012).

C. Matthiesen, A. Ni. Vamivakas, and M. Atatüre, Subnatural linewidth single photons from a quantum dot, Phys. Rev. Lett. 108, 093602 (2012).

N. H. Bonadeo, J. Erland, D. Gammon, D. Park, D. S. Katzer, D. G. Steel, Coherent optical control of the quantum state of a single quantum dot, Science 282, 1473 (1998).

B. Patton, U. Woggon, and W. Langbein, Coherent control and polarization readout of individual excitonic states, Phys. Rev. Lett. 95, 266401 (2005).

H. Htoon, T. Takagahara, D. Kulik, O. Baklenov, A. L. Holmes, Jr. and C. K. Shih, Interplay of Rabi oscillations and quantum interference in semiconductor quantum Dots, Phys. Rev. Lett. 88, 087401 (2002).

T. H. Stievater, Xiaoqin Li, D. G. Steel, D. Gammon, D. S. Katzer, D. Park, C. Piermarocchi, and L. J. Sham, Rabi oscillations of excitons in single quantum dots, Phys. Rev. Lett. 87, 133603 (2001).

Z. Yuan, B. E. Kardynal, R. M. Stevenson, A. J. Shields, C. J. Lobo, K. Cooper, N. S. Beattie, D. A. Ritchie, and M. Pepper, Electrically driven single-photon source, Science 295, 102 (2002).

H. J. Briegel, W. Dür, J. I. Cirac, P. Zoller, Quantum repeaters: The role of imperfect local operations in quantum communication, Phys. Rev. Lett. 81, 5932 (1998).

S. Hughes and C. Roy, Nonlinear photon transport in a semiconductor waveguide-cavity system containing a single quantum dot: Anharmonic cavity-QED regime, Phys. Rev. B 85, 035315 (2012).

C. Van Vlack, P. T. Kristensen, and S. Hughes, Spontaneous emission spectra and quantum light-matter interactions from a strongly coupled quantum dot metal-nanoparticle system, Phys. Rev. B 85, 075303 (2012).

M. Lermer, N. Gregersen, F. Dunzer, S. Reitzenstein, S. Höfling, J. Mørk, L. Worschech, M. Kamp, and A. Forchel, Bloch-wave engineering of quantum dot micropillars for cavity quantum electrodynamics experiments, Phys. Rev. Lett. 108, 057402 (2012).

A. Muller, E. B. Flagg, P. Bianucci, X. Y. Wang, D. G. Deppe, W. Ma, J. Zhang, G. J. Salamo, M. Xiao, and C. K. Shih, Resonance fluorescence from a coherently driven semiconductor quantum dot in a cavity, Phys. Rev. Lett. 99, 187402 (2007).

X. D. Xu, B. Sun, P. R. Berman, D. G. Steel, A. S. Bracker, D. Gammon, and L. J. Sham, Coherent optical spectroscopy of a strongly driven quantum dot, Science 317, 929 (2007).

X. D. Xu, B. Sun, E. D. Kim, K. Smirl, P. R. Berman and D. G. Steel, A. S. Bracker, D. Gammon, and L. J. Sham, Single charged quantum dot in a strong optical field: absorption, gain, and the ac-Stark effect, Phys. Rev. Lett. 101, 227401 (2008).

C. C. Yu, J. R. Bochinski, T. M. V. Kordich, T. W. Mossberg and Z. Ficek, Driving the driven atom: spectral signatures, Phys. Rev. A 56, R4381 (1997).

A. Zrenner, E. Beham, S. Stufler, E Findeis, M. Bichler, and G. Abstreiter, Coherent properties of a two-level system based on a quantum-dot photodiode, Nature (London) 418, 612 (2002).

S. Stufler, P. Ester, A. Zrenner, and M. Bichler, Quantum optical properties of a single $In_xGa_{1-x}As$ − GaAs quantum dot two-level system, Phys. Rev. B 72, 121301(R) (2005).

T. Inoshita and H. Sakaki, Electron relaxation in a quantum dot: Significance of multiphonon processes, Phys. Rev. B 46, 7260 (1992).

I. Wilson-Rae and A. Imamoğlu, Quantum dot cavity-QED in the presence of strong electron-phonon interactions, Phys. Rev. B 65, 235311 (2002).

R. W. Boyd, Nonlinear Optics (Academic Press, Amsterdam) pp. 313 (2008).

C. W. Gardiner and P. Zoller, Quantum Noise (Springer, New York) (2004).

M. O. Scully and M. S. Zubairy, Quantum Optics (Cambridge University Press) (1997).

K. Huang, and A. Rhys, Theory of light absorption and non-radiative transitions in F-Centres, Proc. R. Soc. Lond. A 204, 406 (1950).

S. Weis, R. Rivière, S. Deléglise, E. Gavartin, O. Arcizet, A. Schliesser, and T. J. Kippenberg, Optomechanically induced transparency, Science 330, 1520 (2010).

L. V. Hau, S. E. Harris, Z. Dutton and C. H. Behroozi, Light speed reduction to 17 metres per second in an ultracold atomic gas, Nature 397, 594 (1999).

J. Larson, G. Morigi, and M. Lewenstein, Cold Fermi atomic gases in a pumped optical resonator, Phys. Rev. A 78, 023815 (2008).

S. Gupta, K. L. Moore, K. W. Murch, and D. M. Stamper-Kurn, Cavity nonlinear optics at low photon numbers from collective atomic motion, Phys. Rev. Lett. 99, 213601 (2007).

S. Sauvage, P. Boucaud, T. Brunhes, M. Broquier, C. Crépin, J.-M. Ortega and J. M. Gérard, Dephasing of intersublevel polarizations in InAs/GaAs self-assembled quantum dots, Phys. Rev. B 66, 153312 (2002).

E. A. Zibik, L. R. Wilson, R. P. Green and J-P. R. Wells, Polaron relaxation channel in InAs/GaAs self-assembled quantum dots, Semicond. Sci. Technol 19, 316 (2004).

R. Heitz, I. Mukhametzhanov, O. Stier, A. Madhukar and D. Bimberg, Enhanced polar exciton-LO-phonon interaction in quantum dots, Phys. Rev. Lett. 83, 4654 (1999).

A. Lezama, S. Barreiro, and A. M. Akulshin, Electromagnetically induced absorption, Phys. Rev. A 59, 4732 (1999).

A. H. Safavi-Naeini, T. P. Mayer Alegre, J. Chan, M. Eichenfield, M. Winger, Q. Lin, J. T. Hill, D. E. Chang, and O. Painter, Electromagnetically induced transparency and slow light with optomechanics, Nature 472, 69 (2011).

G. D. Mahan, Many-particle physics, (Plenum, New York, 1981).

R. S. Bennink, R. W. Boyd, C. R. Stroud, and V. Wong, Enhanced self-action effects by electromagnetically induced transparency in the two-level atom, Phys. Rev. A 63, 033804 (2001).

S. E. Harris, J. E. Field, and A. Kasapi, Dispersive properties of electromagnetically induced transparency, Phys. Rev. A 46, R29 (1992).

R. W. Boyd, and D. J. Gauthier, Controlling the velocity of light pulses, Science 326, 1074 (2009).

J. Lu, P. S. E. Yeo, C. K. Gan, P. Wu, and K. P. Loh, Transforming C_{60} molecules into graphene quantum dots, Nat. Nanotech. 6, 247 (2011).

Chapter 6

Nanomechanical Resonator Coupled to a Single Quantum Dot

Due to the limited environment and small size, nanomechanical resonators (NR) have recently attracted considerable interest, as they allow ultrasensitive detections of mass [Lassagne (2008); Jensen (2008); Ekinci (2008); Focus (2011); Li (2011)], mechanical displacements [LaHaye (2004)], and spin [Rugar (2004)]. NRs have been deployed to perform mundane tasks in present technology, such as opening and closing valves, turning mirrors and regulating optical field, etc. Today, numerous companies, from the semiconductor giants to fledgling startups, are making NR based devices for a wide range of consumers [Ekinci (2005)]. Another exciting possibility is to physically couple nanomechanical resonator to a condensed matter system such as semiconductor quantum dots. Such coupled quantum mechanical systems have some application in high precision measurements, ultrasensitive mass sensing and mechanical resonator cooling, and can be used to investigate some basic quantum effects.

In 2004, Wilson-Rae and Imamoğlu [Wilson-Rae (2004)] have demonstrated that it is possible to cool a nanomechanical resonator to its ground state while coupling a single quantum dot. The underlying interactions will also enable quantum state engineering of nonclassical states of motion, including generation and detection of a Fock state and squeezed states of motion [Jelezko (2003)]. In 2010, Bennett et al. [Bennett (2010)] have presented experimental results on mechanical damping of a nanomechanical resonator strongly coupled to a self-assembled InAs quantum dot. They observed highly asymmetric line shapes of Coulomb blockade peaks in the damping that reflect the degeneracy of energy levels on the dot, and predicted the excited state spectroscopy. Recently, Wallraff et al. [Frey (2012)] have experimentally demonstrated the realization of a hybrid solid-state quantum device, in which a semiconductor double quantum dot is dipole coupled to a waveguide resonator. This proposal offers a new way to probe semiconductor quantum systems in the microwave regime, and may be used, for example, for high-energy-resolution measurements of double quantum dots [Li (2011)] and fast time-resolved measurements, in addition to being a promising platform for potentially scalable hybrid solid-state quantum information processing.

Besides, the quality factor ($Q = \omega_n/\gamma_n$) attained to date is an additional and important attribute for nanomechanical resonator. It determines the sensitivity of nanomechanical sensor and insertion loss. For signal processing device, high Q factor directly translates into high sensitivity and low insertion loss [Ekinci (2005)]. In the present chapter, we study the coherent optical spectroscopy of a coupled nanomechanical resonator and a single quantum dot, in the presence of two optical fields. We propose an all-optical method to determine the vibrational frequency and the lifetime of nanomechanical resonator, which directly determine the quality factor of NR [Li (2010a, 2009b); Li and Zhu (2009c); Li (2011d)]. Besides, the coupling strength between quantum dot and nanomechanical resonator is also measured during this process. Although there are other techniques to detect these parameters [Ekinci (2005)], until now this is the unique all-optical idea to realize such measurements. Furthermore, in the last section, we put forward two all-optical devices based on coupled quantum dot and nanomechanical resonator system, i.e., a single photon router and an optical Kerr switch [Li (2010e)].

The physical process we used is called mechanically induced coherent population oscillation (MICPO) [Li (2010a, 2009b)]. As we described before, there are two main techniques to realize the optical transparency, i.e., the electromagnetically induced transparency (EIT) and the coherent population oscillation (CPO). EIT known as coherent population trapping (CPT) [Boller (1991); Harris (1997); Scully (1997); Arimondo (1996)], which constitutes the reduction in resonant light absorption in a very narrow spectral region due to destructive interference of excitation pathways, is one of the most commonly utilized phenomena related to ground-state coherence. Besides, CPO [Bigelow and Lepeshkin (2003a); Bigelow (2003b); Yelleswarapu (2008)] is a process that can lead to transparency and rapid spectral variation of the refractive index. CPO produces a narrow hole in the absorption or gain profile as a consequence of the periodic modulation of the ground-state population at the beat frequency between a strong pump field and a weak probe field sharing a common atomic transition [Boyd (2008)].

However, unlike many other quantum coherence effects, CPO not only can produce a broad transparency spectrum but also is highly insensitive to the presence of dephasing collisions and thus can occur in room-temperature solids [Bigelow (2003b)]. Unlike EIT or CPO described before, the present method describes the resonant interactions of resonator mode and exciton of quantum dot with two optical fields. Since investigating the optical spectrum, we show that the mechanically induced coherent population oscillation can be controlled simply by the intensity of the pump field.

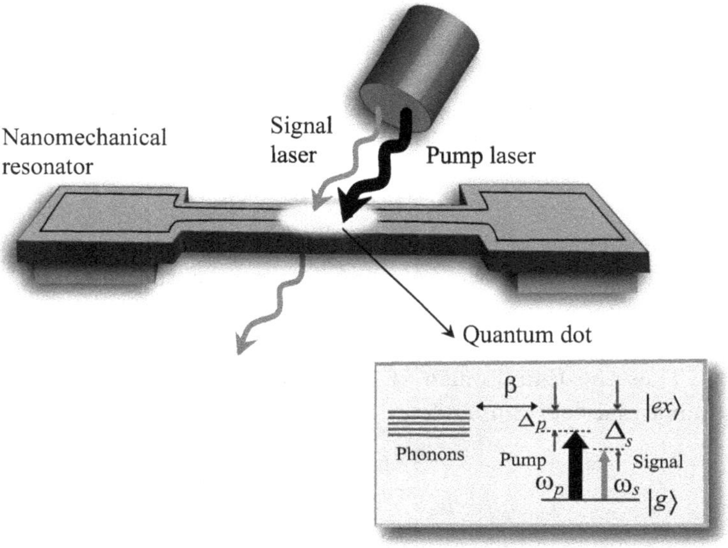

Fig. 6.1 Schematic diagram of a GaAs nanomechanical resonator with an embedded InAs quantum dot in the presence of a strong pump field and a weak signal field. The inset is an energy-level diagram of a quantum dot.

Considering the situation where a semiconductor quantum dot is embedded in the center of a nanomechanical resonator, as shown in Fig.6.1. The quantum dot can be modeled as a two-level system consisting of the ground state $|g\rangle$ and the single exciton state $|ex\rangle$ [Zrenner (2002); Stufler (2005)]. This two-level exciton can be characterized by the pseudospin$-1/2$ operators $S^{\pm}$ and S^z. Then the Hamiltonian describing the quantum dot, the nanomechanical resonator, and the coupling between them is described as follows

$$H_1 = H_{QD} + H_n + H_{QD-n}. \tag{6.1}$$

The first term $H_{QD} = \hbar\omega_{\text{ex}}S^z$ describes the two-level exciton in a quantum dot, where ω_{ex} and $\hbar$ are the exciton frequency and Planck's constant, respectively. The second term $H_n = \hbar\omega_n a^+ a$ represents the fundamental flexural mode of the nanomechanical resonator at frequency ω_n, a and a^+ being bosonic annihilation and creation operators obeying the commutation rules $[a, a^+] = 1$. It should be noted here that in a structure where the thickness of the beam is smaller than its width, and the lowest-energy resonance corresponds to the fundamental flexural mode which will constitute the resonator mode [Wilson-Rae (2004)]. Since this flexion induces extensions and compressions in the structure of Fig.6.1, this longitudinal strain will modify the energy of the electronic states of quantum dot through deformation potential coupling. Then the Hamiltonian

$$H_{QD-n} = \hbar\omega_n \beta S^z(a^+ + a), \tag{6.2}$$

describes the coupling of the resonator mode to the quantum dot [Wilson-Rae (2004)], where β is the coupling strength of the resonator mode-quantum dot.

The optical fields treating classically will interact with quantum dot via extiton. Thus the Hamiltonian of the quantum dot coupling to a pump field and a signal field can be described as [Boyd (2008)]

$$
\begin{aligned}
H_{QD-o} = &-\mu(S^+ E_p e^{-i\omega_p t} + S^- E_p^* e^{i\omega_p t}) \\
&-\mu(S^+ E_s e^{-i\omega_s t} + S^- E_s^* e^{i\omega_p t}),
\end{aligned} \tag{6.3}
$$

where μ is the electric dipole moment of the exciton, ω_p (ω_s) is the frequency of the pump field (signal field), and E_p (E_s) is the slowly varying envelope of the pump field (signal field). Most recently, Xu *et al.* [Xu (2007, 2008); Xu and Sun (2008)] have experimentally observed the coherent optical spectroscopy in a strongly driven quantum dot without a nanomechanical resonator.

Therefore, the total Hamiltonian of the coupled quantum dot-nanomechanical resonator in the presence of two optical fields is given by [Wilson-Rae (2004); Boyd (2008)]

$$
\begin{aligned}
H = &\, H_1 + H_{QD-o} \\
= &\, \hbar\omega_{\text{ex}}S^z + \hbar\omega_n a^+ a + \hbar\omega_n \beta S^z(a^+ + a) \\
&- \mu(S^+ E_p e^{-i\omega_p t} + S^- E_p^* e^{i\omega_p t}) - \mu(S^+ E_s e^{-i\omega_s t} + S^- E_s^* e^{i\omega_s t}). \tag{6.4}
\end{aligned}
$$

In a rotating frame at the pump field frequency ω_p, the total Hamiltonian of the system reads as follows

$$H = \hbar\Delta_p S^z + \hbar\omega_n a^+ a + \hbar\omega_n \beta S^z(a^+ + a) - \hbar(\Omega S^+ + \Omega^* S^-)$$
$$-\mu(S^+ E_s e^{-i\delta t} + S^- E_s^* e^{i\delta t}), \tag{6.5}$$

where $\Delta_p = \omega_{\mathrm{ex}} - \omega_p$ and $\delta = \omega_s - \omega_p$ are the exciton-pump frequency detuning and signal-pump detuning, respectively. $\Omega = \mu E_p/\hbar$ is the Rabi frequency of the pump field.

In terms of the Heisenberg equation of motion $i\hbar dO/dt = [O, H]$, and the commutation relation $[S^z, S^\pm] = \pm S^\pm$, $[S^+, S^-] = 2S^z$, we obtain the temporal evolutions of the exciton and nanomechanical resonator in the coupled quantum dot-nanomechanical resonator system as follows

$$\frac{dS^z}{dt} = i(\Omega S^+ - \Omega S^-) + \frac{i\mu E_s e^{-i\delta t}}{\hbar}S^+ - \frac{i\mu E_s^* e^{i\delta t}}{\hbar}S^-, \tag{6.6}$$

$$\frac{dS^-}{dt} = -i\Delta_p S^- - i\omega_n \beta N S^- - 2i\Omega S^z - \frac{2i\mu E_s e^{-i\delta t}}{\hbar}S^z, \tag{6.7}$$

$$\frac{d^2 N}{dt^2} + \omega_n^2 N + 2\omega_n^2 \beta S^z = 0, \tag{6.8}$$

where $N = a^+ + a$. In what follows, we ignore the quantum properties of S^z, S^- and N [Lam (1991); Zhu (2001)], and then the semiclassical equations read as follows

$$\frac{dS^z}{dt} = -\Gamma_1(S^z + 1/2) + i\Omega S^+ - i\Omega S^- + \frac{i\mu E_s e^{-i\delta t}}{\hbar}S^+ - \frac{i\mu E_s^* e^{i\delta t}}{\hbar}S^-, \tag{6.9}$$

$$\frac{dS^-}{dt} = -(i\Delta_p + \Gamma_2)S^- - i\omega_n \beta N S^- - 2i\Omega S^z - \frac{2i\mu E_s e^{-i\delta t}}{\hbar}S^z, \tag{6.10}$$

$$\frac{d^2 N}{dt^2} + \gamma_n \frac{dN}{dt} + \omega_n^2 N = -2\omega_n^2 \beta S^z, \tag{6.11}$$

where we introduced the damping terms phenomenologically [Boyd (2008)], so here Γ_1 and Γ_2 are the exciton relaxation rate and dephasing rate, respectively. γ_n is the decay rate of the nanomechanical resonator due to the coupling to a reservoir of "background" modes and the other intrinsic processes [Wilson-Rae (2004); Ekinci (2005); Wilson-Rae (2008)]. In order to solve Eqs.(6.9)-(6.11), we make the ansatz [Boyd (2008)]: $S^z(t) = S_0^z + S_+^z e^{-i\delta t} + S_-^z e^{i\delta t}$, $S^-(t) = S_0 + S_+ e^{-i\delta t} + S_- e^{i\delta t}$, $N(t) = N_0 + N_+ e^{-i\delta t} + N_- e^{i\delta t}$. On substituting them in Eqs.(6.9)-(6.11) and on working to the lowest order in E_s, but to all orders in E_p , we obtain in the steady state:

$$- (i\Delta_p + \Gamma_2)S_0 - i\omega_n\beta N_0 S_0 - 2i\Omega S_0^z = 0, \tag{6.12}$$

$$- (i\Delta_p + \Gamma_2 - i\delta)S_+ - i\omega_n\beta(N_0 S_+ + N_+ S_0) - 2i\Omega S_+^z = \frac{2i\mu E_s S_0^z}{\hbar}, \tag{6.13}$$

$$- (i\Delta_p + \Gamma_2 + i\delta)S_- - i\omega_n\beta(N_0 S_- + N_- S_0) - 2i\Omega S_-^z = 0, \tag{6.14}$$

$$- \frac{\Gamma_1}{2} - \Gamma_1 S_0^z + i\Omega S_0^* - i\Omega S_0 = 0, \tag{6.15}$$

$$- \Gamma_1 S_+^z + i\Omega S_-^* - i\Omega S_+ + \frac{i\mu E_s S_0^*}{\hbar} = -i\delta S_+^z, \tag{6.16}$$

$$- \Gamma_1 S_-^z + i\Omega S_+^* - i\Omega S_- - \frac{i\mu E_s^* S_0}{\hbar} = i\delta S_-^z, \tag{6.17}$$

$$\omega_n^2 N_0 = -2\omega_n^2\beta S_0^z, \tag{6.18}$$

$$(-\delta^2 - i\gamma_n\delta + \omega_n^2)N_+ = -2\beta\omega_n^2 S_+^z, \tag{6.19}$$

$$(-\delta^2 + i\gamma_n\delta + \omega_n^2)N_- = -2\beta\omega_n^2 S_-^z. \tag{6.20}$$

From the solutions of Eqs.(6.12)-(6.20), we can obtain S_+ which corresponds to the linear optical susceptibility as follows

$$\chi^{(1)}(\omega_s) = \frac{\mu S_+}{E_s} = \frac{\mu^2}{\Gamma_2\hbar}\chi(\omega_s), \tag{6.21}$$

where the dimensionless susceptibility is given by

$$\chi^{(1)}(\omega_s) = \frac{2bw_0(\Omega_R^2 + c) - ew_0}{ae - 2b(\Omega_R^2 + c)(b - \delta_0)}, \tag{6.22}$$

where $a = \Delta_{p0} - \omega_{n0}\beta^2 w_0 - i - \delta_0$, $b = \Delta_{p0} - \omega_{n0}\beta^2 w_0 + i + \delta_0$, $c = \Omega_R^2\omega_{n0}\beta^2\eta w_0/(\Delta_{p0} - \omega_{n0}\beta^2 w_0 - i)$, $d = \Omega_R^2\omega_{n0}\beta^2\eta w_0/(\Delta_{p0} - \omega_{n0}\beta^2 w_0 + i)$, $e = (2\Omega_R^2 + 2d - 2ib - b\delta_0)(b - \delta_0)$, $\eta = \omega_{n0}^2/(\omega_{n0}^2 - \delta_0^2 - i\delta_0\gamma_{n0})$, and $w_0 = 2S_0^z$, $\delta_0 = \delta/\Gamma_2$, $\Omega_R = \Omega/\Gamma_2$, $\omega_{n0} = \omega_n/\Gamma_2$, $\Delta_{p0} = \Delta_p/\Gamma_2$, $\gamma_{n0} = \gamma_n/\Gamma_2$, $\Gamma_1 = 2\Gamma_2$.

In the same way, the nonlinear optical susceptibility can be obtained by

$$\chi_{\text{eff}}^{(3)}(\omega_s) = \frac{\mu S_-}{3E_p^2 E_s^*} = \frac{\mu^4}{3\hbar^3\Gamma_2^3}\chi^{(3)}(\omega_s) = \Sigma_3\chi^{(3)}(\omega_s), \tag{6.23}$$

where $\Sigma_3 = \mu^4/(3\hbar^3\Gamma_2^3)$ and the dimensionless nonlinear optical susceptibility is given by

$$\chi^{(3)}(\omega_s) = \frac{w_0(A - \delta_0) - Bw_0}{AE\Omega_R^2 - B\Omega_R^2(A - \delta_0)}, \qquad (6.24)$$

where $A = \Delta_{p0} - \beta^2\omega_{n0}w_0 - i + \delta_0$, $B = \Delta_{p0} - \beta^2\omega_{n0}w_0 + i - \delta_0$, $C = \Omega_R^2\omega_{n0}\beta^2\xi w_0/(\Delta_{p0} - \omega_{n0}\beta^2 w_0 - i)$, $D = \Omega_R^2\omega_{n0}\beta^2\xi w_0/(\Delta_{p0} - \omega_{n0}\beta^2 w_0 + i)$, $E = (2\Omega_R^2 + 2D - 2iB + B\delta_0)(A - \delta_0)/(2C + 2\Omega_R^2)$, and $\xi = \omega_{n0}^2/(\omega_{n0}^2 - \delta_0^2 + i\delta_0\gamma_{n0})$.

The population inversion w_0 of the exciton is determined by

$$(w_0 + 1)[(\Delta_{p0} - \beta^2\omega_{n0}w_0)^2 + 1] = -2\Omega_R^2 w_0. \qquad (6.25)$$

It should be noticed that Eq.(6.25) has the same form with Eq.(6.17) in
Chapter 5. Figure 6.2 shows the steady-state value of the population inversion w_0 as a function of the pump-exciton detuning (a) when fixing
values of pump Rabi frequency, and (b) as a function of the pump Rabi
frequency when fixing value of the pump-exciton detuning. The bistable

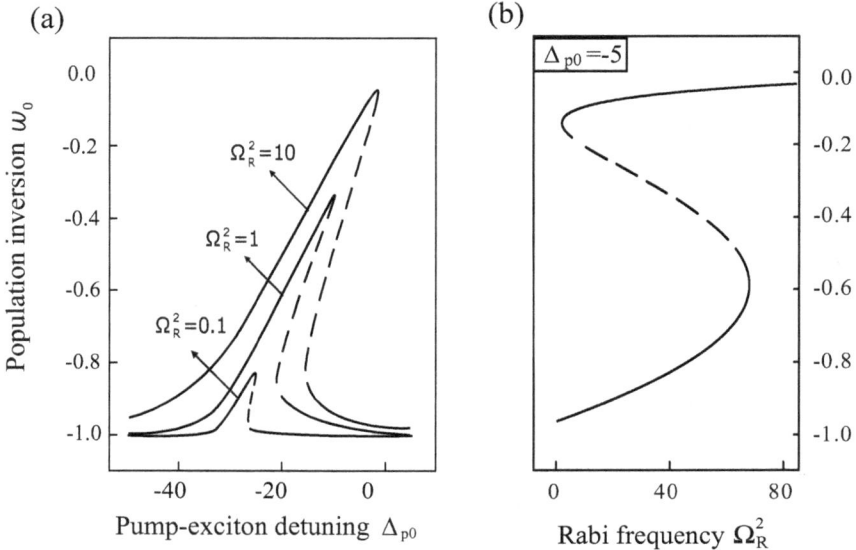

Fig. 6.2 (a) Steady-state population inversion w_0 as a function of pump-exciton detuning with different Rabi frequencies of the pump field. The dashed curves and the solid
curves correspond to the unstable states and table state, respectively. (b) Population
inversion w_0 as a function of pump Rabi frequency. The parameters used are $\beta = 0.6$
and $\omega_{n0} = 50$.

behavior in Fig.6.2 provides a strong evidence that the coupled quantum dot and nanomechanical resonator system can really act as an optomechanical system [Brennecke (2008); Kanamoto (2010)], where the role of the optical cavity is played by the excitonic resonance of the QD, and the role of the mechanical element is played by the vibrations of the NR. We also notice that the value of population inversion w_0, confined between 0 and -1, corresponds to the independence between deformation coupling of exciton-phonon and pump field, which is very similar with a single quantum dot system proposed in Chapter 5 and is also different from the cavity optomechanical system.

6.1 Mechanically induced coherent population oscillation (MICPO)

For illustration of the numerical results, we choose the realistic system of an InAs quantum dot coupled to a GaAs nanomechanical resonator. The parameters used are $(\beta, \Gamma_1, \omega_n, \gamma_n) = (0.06, 0.3\text{GHz}, 1.2\text{GHz}, 4.0 \times 10^{-5}\text{GHz})$ [Wilson-Rae (2004)].

Figure 6.3(a) shows the signal absorption spectrum as a function of detuning Δ_s, where three main features are obvious, i.e., the middle small peaks, the left negative absorption peak and the right absorption peak. In analogy with Fig.5.5 in Chapter 5, the three features attribute to the vibration of nanomechanical resonator, which can be called mechanically induced coherent population oscillation (MICPO). Figure 6.3(b) gives the origin of these new features. Part (1) shows the dressed states of exciton when dressing with the vibration mode of nanomechanical resonator. In part (2), the electrons make a transition between the lowest energy state $|g, n\rangle$ and the highest energy state $|\text{ex}, n+1\rangle$. Since absorption of two pump photons and emission of a photon at $\omega_p - \omega_n$, this process can be treated as mechanically induced three-photon resonance. This process can amplify a wave at $\delta = -\omega_n = -1.2\text{GHz}$, as indicated by the region of negative absorption in Fig.6.3(a). Part (3) shows the origin of mechanically induced stimulated Rayleigh resonance, which corresponds to a transition $|g, n\rangle \rightarrow |\text{ex}, n\rangle$. Each of these transitions is centered on the frequency of the pump field. The rightmost part (4) corresponds to the mechanically induced absorption resonance as modified by the ac Stark effect.

Figure 6.4 plots the imaginary part of the linear optical susceptibility $\chi^{(1)}$ with different Rabi frequencies of pump field, in the presence of a weak

(a)

(b)

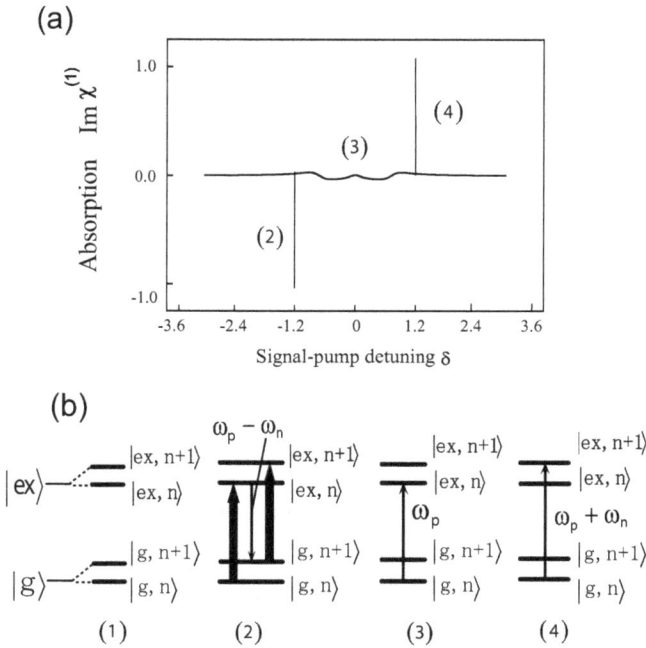

Fig. 6.3 (a) In the presence of a strong pump field, the absorption spectrum of a signal field as a function of signal-pump detuning. The parameters used are $\Omega^2 = 0.2(\text{GHz})^2$, $\omega_n = 1.2\text{GHz}$, $\Delta_p = 0$, $\gamma_n = 4 \times 10^{-5}\text{GHz}$, and $\beta = 0.06$. (b) The new features in the spectrum shown in (a) are identified by the corresponding transition between the dressed states of exciton. $|n\rangle$ denotes the number states of the nanomechanical resonator.

signal field and a strong pump field. It should be noticed that the signal field become transparent at $\Delta_s = 0$, in the presence of pump field. Different pump intensities result in different absorption rates. However, if removing the pump field, the solid curve in Fig.6.4 indicates the normal absorption of the excitonic transition. Therefore, the MICPO induced transparency can be controlled continuously by adjusting the Rabi frequency of pump field. The coupling QD-NR system leads to transparency in essentially the same way as the three-level systems [Harris (1997); Scully (1997); Arimondo (1996)].

Figure 6.5 discusses the role of coupling strength between NR and QD during the transparency process. The solid curve in Fig.6.5 shows that even there is a strong pump field, the transparency phenomenon of signal field will disappear immediately when the coupling strength $\beta = 0$. The other curves in Fig.6.5 give us different coupling strengths related to different

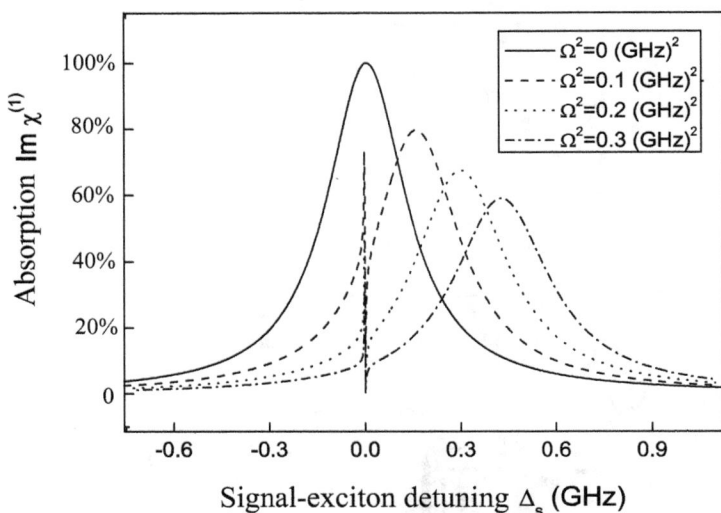

Fig. 6.4 The dimensionless imaginary part of the linear optical susceptibility as a function of the detuning Δ_s for different Rabi frequencies. The other parameters used are $\omega_n = 1.2\text{GHz}$, $\Delta_p = 1.2\text{GHz}$, $\gamma_n = 4.0 \times 10^{-5}\text{GHz}$, and $\alpha = 0.06$.

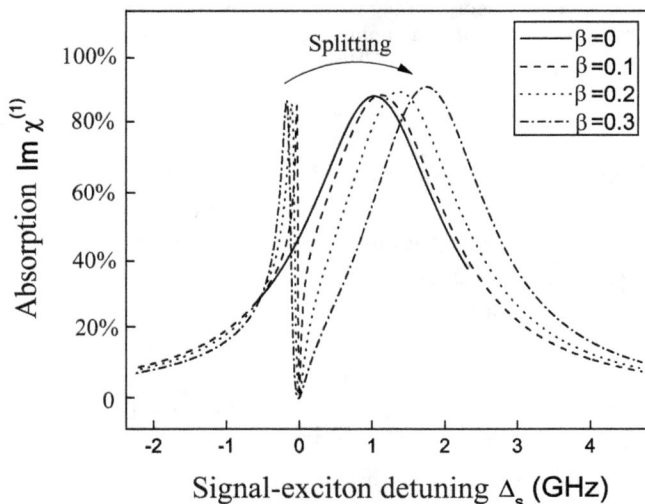

Fig. 6.5 The absorption spectrum of a signal field as a function of the detuning Δ_s for four different coupling strengths. The parameters used are $\Omega^2 = 0.1(\text{GHz})^2$, $\omega_n = 1.2\text{GHz}$, $\Delta_p = 1.2\text{GHz}$, and $\gamma_n = 4.0 \times 10^{-5}\text{GHz}$.

materials, which all result in transparency at $\Delta_s = 0$. The larger the coupling strength, the longer the distance between two split peaks.

Figure 6.6 shows the relationship between Δ_p and transparency effect, due to mechanically induced coherent population oscillation. It should be noticed that the location of transparency window satisfies $\Delta_p + \Delta_s = \omega_n = 1.2$GHz, i.e., if setting $\Delta_p = 1.2$GHz, the absorption spectrum will leads to a transparency at $\Delta_s = 0$ (the full curve); and if $\Delta_p = 0.8$GHz, the deep hole will be at $\Delta_s = \omega_n - \Delta_p = 0.4$GHz (the dotted curve). In this case, the location of the MICPO window can be modulated by Δ_p, which shows the quantized process of MICPO.

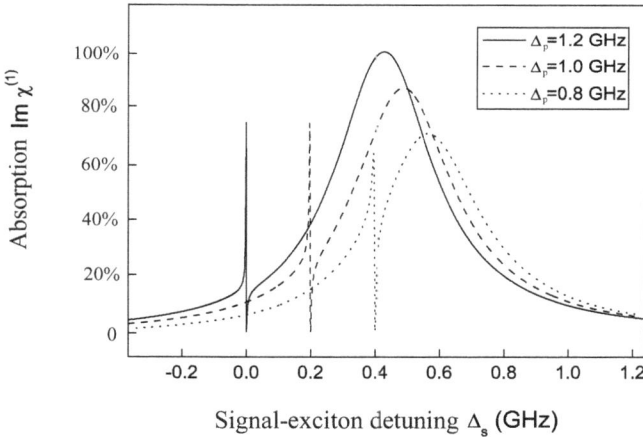

Fig. 6.6   The linear optical susceptibility of signal field as a function of the detuning Δ_s for different detunings of Δ_p. The parameters used are $\Omega^2 = 0.1(\text{GHz})^2$, $\omega_n = 1.2$GHz, $\Delta_p = 0$, and $\gamma_n = 4.0 \times 10^{-5}$GHz.

6.2   Measurement of vibrational frequency of NR

In Fig.6.7, we propose a method to measure the vibrational frequency of nanomechanical resonator, according to MICPO. It should be noticed that the two steep peaks at the both sides of signal spectrum correspond to the vibrational frequency of nanomechanical resonator, i.e., when $\omega_n = 1.2$GHz, the two steep peak will at $\Delta_s = \pm 1.2$GHz, respectively (the bottom curve in Fig.6.7); when $\omega_n = 1.0$GHz, the two steep peak will at $\Delta_s = \pm 1.0$GHz (the middle curve in Fig.6.7). In this case, if fixing the pump field on the resonance of exction frequency ($\Delta_p = 0$) and scanning the second

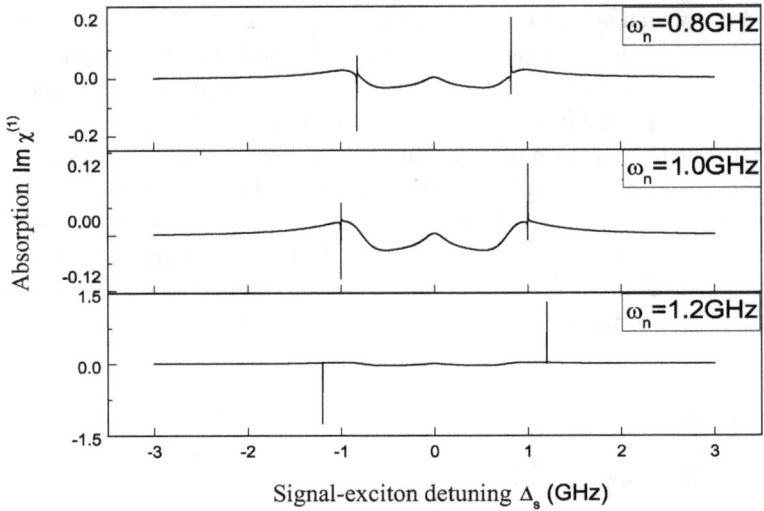

Fig. 6.7   The signal absorption spectrum as a function of signal-exciton detuning, with different vibrational frequencies of nanomechanical resonator. The parameters used are the same with Fig.6.5.

signal across the exciton frequency, the two steep peaks appearing at the signal absorption spectrum will just relate to the vibrational frequency of nanomechanical resonator. The frequency measurement of nanomechanical resonator proposed here is the first all-optical technique, which will enhance the measurement sensitivity in the future.

6.3   Measurement of coupling strength between NR and QD

In Fig.6.5, we notice that the speak splitting is changed with the different coupling strength between nanomechanical resonator and quantum dot. In this case, we plot Fig.6.8 which shows the proportional relationship between speak splitting and coupling strength. If we fix the pump-exciton detuning on the frequency of nanomechanical resonator $(\Delta_p = \omega_n)$, and scan the second signal field across the exciton frequency, the distance between two peaks around the transparency window of signal absorption spectrum will be proportional to the coupling strength between NR and QD. If the distance can be detected, we can know the value of coupling strength.

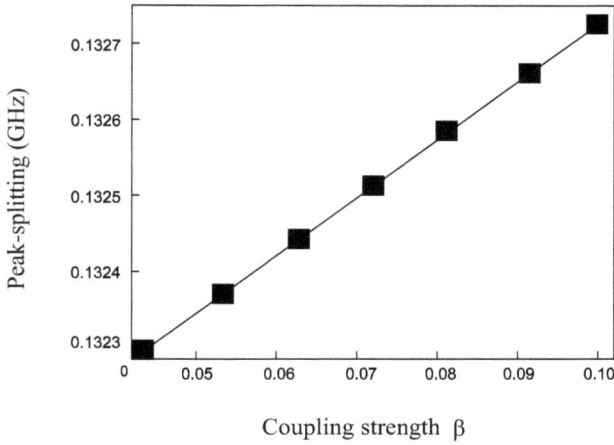

Fig. 6.8   The relationship between peak splitting in Fig.6.5 and coupling strength. The parameters used are the same with Fig.6.5.

6.4   Measurement of lifetime of NR

Figure 6.9 demonstrates the behavior of the lifetime of nanomechanical resonator. It is obvious that small decay rate yields the extremely sharp discrete resonant peaks. Thanks to the long lifetime of resonator, we not only generate the transparency in this coupling system, but also obtain the two sharp peaks at the both sides of signal spectrum. Here, the lifetime (τ_n) is inversely related to the decay rate of nanomechanical resonator (γ_n). From the figure we can see that increasing the value of decay rate, the two peaks will become smaller. Compared with the two peaks in Fig.5.7 showing the frequency of LO-phonon in a single quantum dot, here the strength of the two steep peaks is obvious stronger, due to the long vibrational lifetime of nanomechanical resonator. The vibrational lifetime of NR is $\tau_n = 1/\gamma_n = 25\mu s$, which is four order of magnitude longer than the dephasing time of exciton ($1/\Gamma_2 = 6.7$ns). In Fig.6.9(a), we hardly identify the two peaks as γ_n turns to 4.0×10^{-3}GHz. Therefore, we conclude that the necessary condition of the realization of mechanically induced coherent population oscillation is the lifetime of mechanical resonator (or the LO-phonon in Chapter 5) is at least one order of magnitude longer than the dephsing time of exciton.

For real nanomechanical resonator, the transparency based on MICPO can be controlled by the lifetime of resonator. We plot the relationship

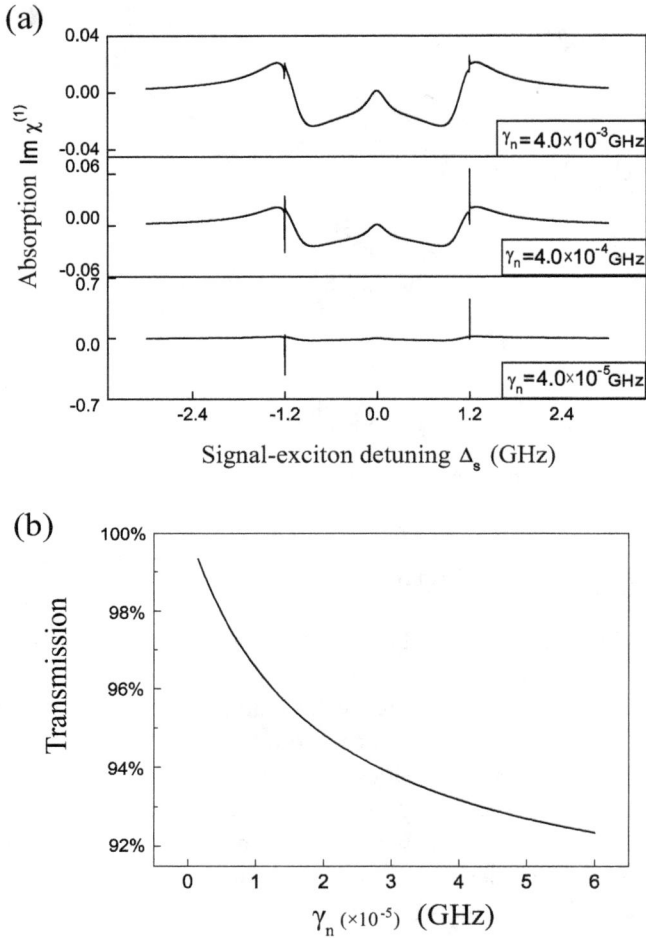

Fig. 6.9   (a) The dimensionless imaginary part of the linear optical susceptibility as a function of the detuning Δ_s for different lifetimes of nanomechanical resonator. (b) The transmission of a signal field as a function of the lifetime of nanomechanical resonator. The parameters used are $\omega_n = 1.2\text{GHz}$, $\Delta_p = 0$, $\Omega^2 = 0.3(\text{GHz})^2$, and $\alpha = 0.06$.

between the transmission and the lifetime of nanomechanical resonator in Fig.6.9(b). It shows that the transmission of the signal beam decreases with increasing of the decay rate of resonator. This picture provides us a simple method to measure the decay rate of nanomechanical resonator. Since measuring the vibrational frequency and decay rate of nanomechanical resonator, the quality factor of NR can be obtained according to $Q = \omega_n/\gamma_n$.

6.5 A single photon router

Recently quantum information science has been developed rapidly due to the substitution of photons as signal carriers rather than the limited electrons [Kimble (2008)]. Designing an optical switch or a photon router oper-

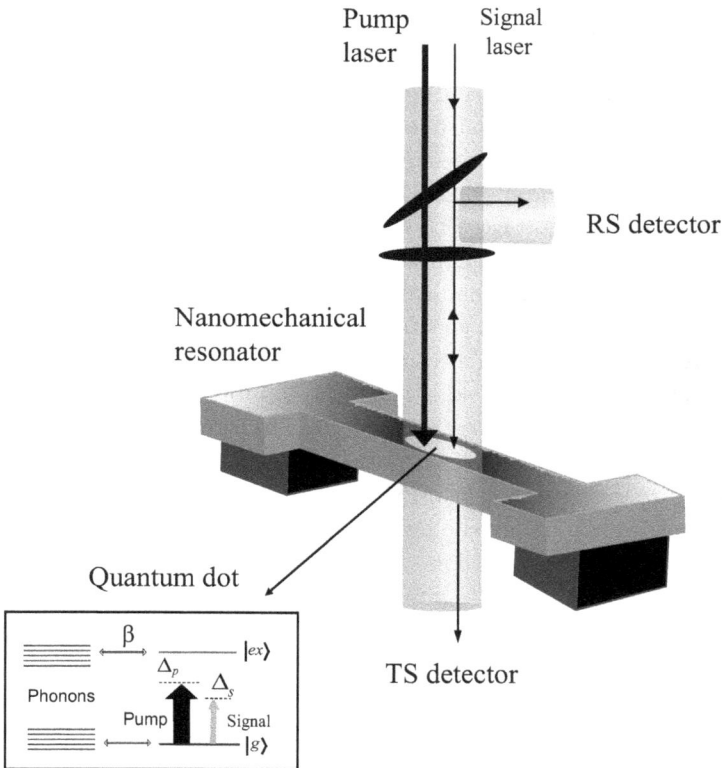

Fig. 6.10 Schematic diagram of a single photon router with a InAs quantum dot embedded in a GaAs nanomechanical resonator in the presence of a strong pump field and a weak signal field. The signal laser is in the single-photon regime, while the pump field is a strong laser. The reflected signal (RS) and transmitted signal (TS) are detected by the specific apparatus. The left bottom window shows the energy levels of exciton in quantum dot while dressing with the vibration modes (phonon modes) of nanomechanical resonator.

ated at a single photon level enables a selective quantum channel in quantum information and quantum networks [Hall (2011); Hoi (2011); Agarwal (2012)].

In a pioneering work, Wilson-Rae *et al.* [Wilson-Rae (2004)] have first theoretically proposed a scheme to cool a nanomechanical resonator via a single quantum dot. After that, we further investigated the light propagation properties of this coupled system, and proposed some all-optical methods to measure the system's parameters and realized some all-optical devices theoretically [Li (2011, 2010a, 2009b); Li and Zhu (2009c); Li (2010e, 2011d)]. Here, in this subsection, we propose a scheme for implementing a single-photon router based on such a single quantum dot coupled to a nanomechanical resonator which operates in the optical regime and at ultralow pump power. The routing principle is as follows. In the absence of the pump field, the signal photons would be reflected; in the presence of a suitable pump field, the signal photons are transmitted due to the MICPO effect. Therefore, we can apply a tunable pump field to switch the route of the signal photons. The single photon router is shown in Fig.6.10, where a strong pump field and a weak signal field are aimed for this coupled quantum dot and nanomechanical resonator system.

The signal transmission spectra and reflection spectra as a function of the signal-exciton detuning, are shown in Fig.6.11, where $\Delta_s = \omega_s - \omega_{ex}$. In the absence of the pump field, Fig.6.11(a) exhibits a standard Lorentzian shape and an inverted Lorentzian shape in the reflection (dashed curve) and transmission spectra (solid curve) of the signal field, which signifies the completely reflected signal beam through the coupled NR-QD. It is clear that no matter how strong the interaction between NR and QD, the signal field is totally reflected without the pump field. However, as the pump beam turns on and fixes the pump-exciton detuning on the red-sideband $\Delta_p = \omega_n$, the reflection spectrum and the transmission spectrum present a completely different behavior as shown in Fig.6.11(b). Here, the transmission is nearly 100% but the reflection is zero at the resonance. Which means, when radiating a pump field on the coupled QD-NR system, the resonant signal beam will transmit the system completely while the reflection is totally suppressed, which is opposite to the case in Fig.6.11(a). In order to clarify the relationship between the pump power and the magnitude of signal transmission, we plot Fig.6.11(c) with $\omega_s = \omega_{ex}$. Figure 6.11(c) shows that when turning off the pump field, the photons are totally reflected. Increasing the pump power results in decreased reflection and increased transmission of the signal field.

(a) Pump off

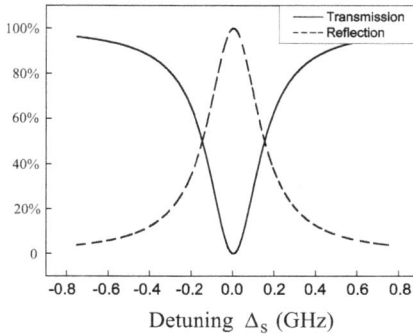

(b) Pump on

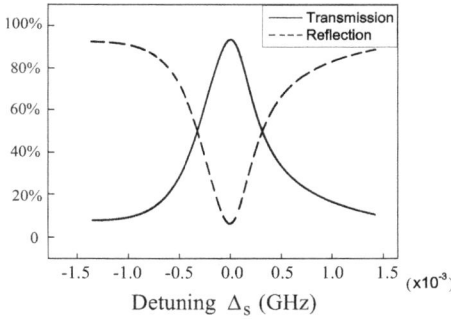

(c) Pump on

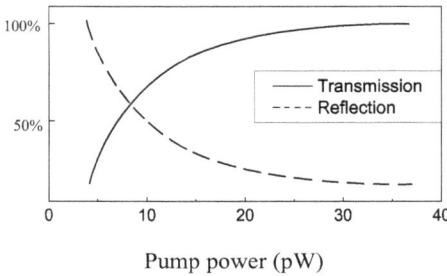

Fig. 6.11   The transmission and reflection spectrum of a signal beam with and without the strong pump field at (a) the pump off situation; (b) the pump on situation where the pump field is on the red-sideband and the pump power is 25pW. (c) The magnitude of transmission and reflection as a function of pump power when the signal field is resonant with the exciton. The other parameters used are $\omega_n = 1.2\text{GHz}$, $\Delta_p = \omega_n = 1.2\text{GHz}$, $\tau_n = 25\mu s$, and $\beta = 0.06$.

For intermediate values of pump power, a part of power will be lost due to the decay of the exciton in the QD. However, the signal field can be transmitted totally as the pump power reaches 30pW. This extremely low pump power enables the coupled quantum dot-nanomechanical resonator system act as a single-photon router in the optical regime, due to the quantum interference between the vibration modes (phonon modes) and the beat of the two optical fields via the quantum dot. If the beat frequency of two lasers $\delta = \omega_s - \omega_p$ is close to the resonance frequency ω_n of the NR, the nanomechanical resonator starts to oscillate coherently, which results in Stokes ($\omega_s = \omega_p - \omega_n$) and anti-Stokes ($\omega_{as} = \omega_p + \omega_n$) scattering of light from the NR via the exciton in the quantum dot. The Stokes scattering is strongly suppressed because it is highly off-resonant with the exciton frequency. However, the anti-Stokes field can interfere with the near-resonant signal beam and thus modify the signal beam spectrum.

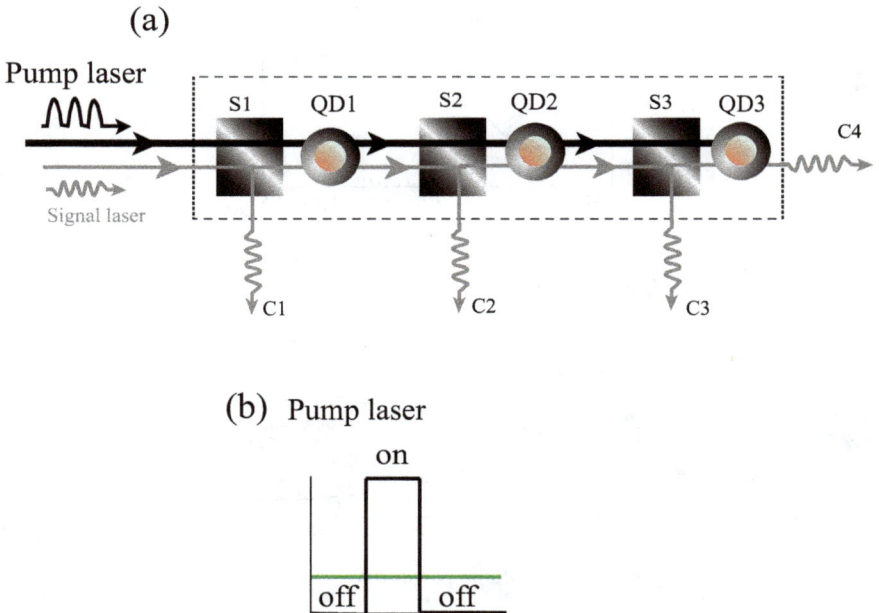

Fig. 6.12 (a) The operation scheme for multi-output router channels, where a four-port router using three quantum dots, labeled as QD1, QD2, and QD3. When inputting a weak signal filed in the single photon regime at ω_s, the output channels (C1, C2, C3, C4) who can receive the signal is decided by the status of pump field (on or off). (b) The pump pulse sequence.

Figure 6.12 plots the scheme of a single photon router, where three quantum dots, labeled as QD1, QD2, and QD3 serve as three channels for optical signal. Clearly, we can embed three quantum dots into one nanomechanical resonator. When fixing the pump laser on the red sideband, i.e., $\Delta_p = \omega_n$, and imputing a weak, resonant and single photon signal to the QD, the signal photon will transmit or reflect from the QD to the TS detector or the RS detector, that depends on the status of pump field (Fig.6.12(b)). Turning on or off the pump pulse result in the transmitted or reflected signal, respectively. Here, we note that such a single photon router works well in the optical regime instead of the microwave regime [Hoi (2011)] and at an ultralow power (25pW) of the pump field, which is required for a practical router used in the quantum information networks. Besides, this single photon router has a short switching time about a few ns for InAs QD, which is determined by the exciton dephasing time of quantum dot. Therefore, this solid-state router can also operate at high speed and high efficiency.

6.6 All-optical Kerr switch

With the rapid development of optics, it is a practical need to understand material's nonlinear behavior in order to avoid it when it is unwanted, and exploit it efficiently when it is wanted [Furuya (2009); Hashemi (2009); Lifshitz (2008)]. At the same time, advances in the fabrication, transduction, and detection of nanomechanical systems, such as cantilever, carbon nanotube and cavity resonator have opened up an exciting new experimental window into the study of fundamental questions in nonlinear effects [Westra (2010); Hendry (2010); Yao (2010)]. In this subsection, we theoretically realize an optical Kerr switch based on a coupled quantum dot and nanomechanical resonator system. Since John Kerr, a Scottish physicist, discovered the Kerr effect in 1875 [Weinberger (2008)], nonlinear Kerr effect has attracted many attentions in different materials. For optical Kerr effect, the electric field of light radiating on a transparent material without absorption will produce an anisotropic refractive index, which is detected by using a second light beam. The lasers pulsed on femtosecond time scales is strong enough to generate a measurable Kerr effect. Furthermore, the optical Kerr effect can be used to study the interactions of condensed systems and molecular structures, i.e., plastic crystals, liquid solutions, pure liquids, etc. [Righini (1993); Saltiel (2005); Li (2005); Zhong (2010)]. Here, the optical

Kerr effect of signal light in a coupled quantum dot and nanomechanical resonator system is achieved by adjusting another strong pump field. Distinct with the above sections (measurement of parameters and a single photon router) where the pump field is on the red sideband, i.e., $\Delta_p = \omega_n$, here the optical Kerr switch is operated on the blue sideband of pump field ($\Delta_p = -\omega_n$).

The real part and the imaginary part of $\chi_{\text{eff}}^{(3)}$ in Eq.(6.23) correspond to the Kerr effect and the nonlinear absorption, respectively [Boyd (2008)]. The all-optical Kerr effect with and without the coupling strength between nanomechanical resonator and quantum dot is shown in Fig.6.13(a). We notice that the value of Kerr coefficient become zero and large when the coupling strength $\beta = 0$ and $\beta \neq 0$, respectively, which means the coupling strength plays an important role in this process. Increasing the pump intensity results in the enhancement of the optical Kerr effect, as shown in Fig.6.13(b). Different from the above sections operating at the red sideband $\Delta_p = \omega_n$, here the all-optical Kerr effect is working at the blue sideband of the pump field $\Delta_p = -\omega_n$. The physical origin of this result is due to mechanically induced coherent population oscillation (MICPO), which makes quantum interference between the resonator and the beat of the two optical fields via the quantum dot when the pump-signal detuning δ is equal to the resonator frequency ω_n. Fig.6.13(c) plot the relationship between the optical Kerr coefficient and the Rabi frequency of pump field. Clearly, the magnitude of Kerr effect can be modulated by the pump field.

Figure 6.14 plots the Kerr coefficient ($\text{Re}\chi_{\text{eff}}^{(3)}$) and the nonlinear optical absorption ($\text{Im}\chi_{\text{eff}}^{(3)}$) as a function of the detuning of the signal field from the exciton resonance Δ_s. It is clear that, at $\Delta_s = 0$ the large enhancement of the Kerr coefficient can be achieved in combination with a negative nonlinear absorption (Fig.6.14(a)) and a vanishing linear absorption (Fig.6.14(b)) simultaneously, with the detuning $\Delta_s = -\omega_n = -1.2\text{GHz}$. In this case, if fixing the signal beam on exciton frequency and turning on the pump field to $\Delta_p = -\omega_n$, we can easily obtain the largely enhanced optical Kerr effect in the signal spectrum immediately. Particularly, for $\Omega^2 = 0.3(\text{GHz})^2$, the Kerr coefficient $\text{Re}\chi_{\text{eff}}^{(3)}(\omega_s) = 1.4 \times 10^{-17}\text{m}^2/\text{V}^2$, provided that the number density of QD $\mathcal{N} = 10^9/\text{m}^3$ and $\mu = 28$ Debye [Jiang (2008); Vamivakas (2009)]. This value is much larger than that in atomic systems [Boyd (2008)].

In view of the above discussions, by fixing the pump-exciton detuning $\Delta_p = 0$, the Kerr coefficient turns to a different behavior, as shown

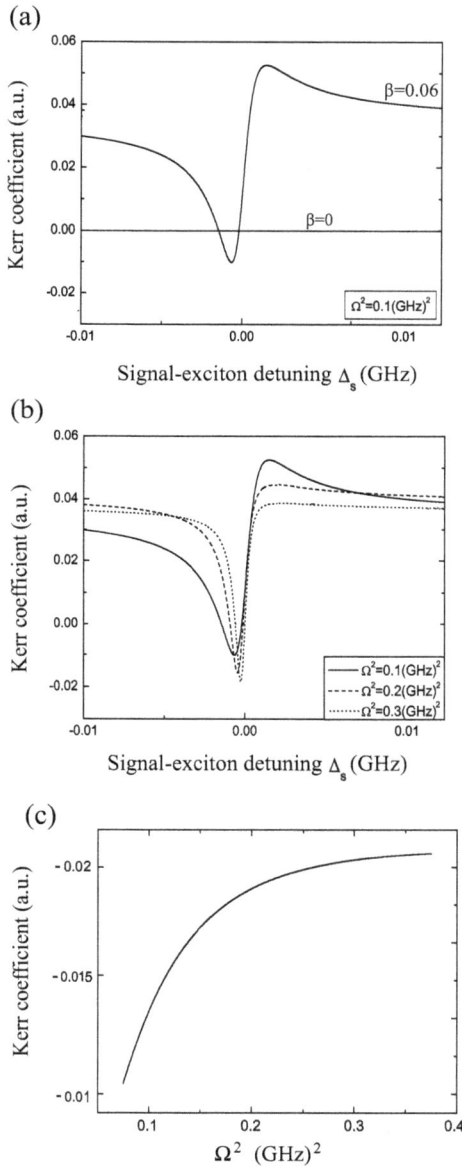

Fig. 6.13 The optical Kerr coefficient as a function of the detuning Δ_s for (a) with and without the coupling strength between NR and QD; (2) different Rabi frequencies of pump field. (c) The magnitude of Kerr coefficient as a function of Rabi frequency of pump field. The parameters used are $\Delta_p = -\omega_n = -1.2\text{GHz}$ and $\gamma_n = 4 \times 10^{-5}\text{GHz}$.

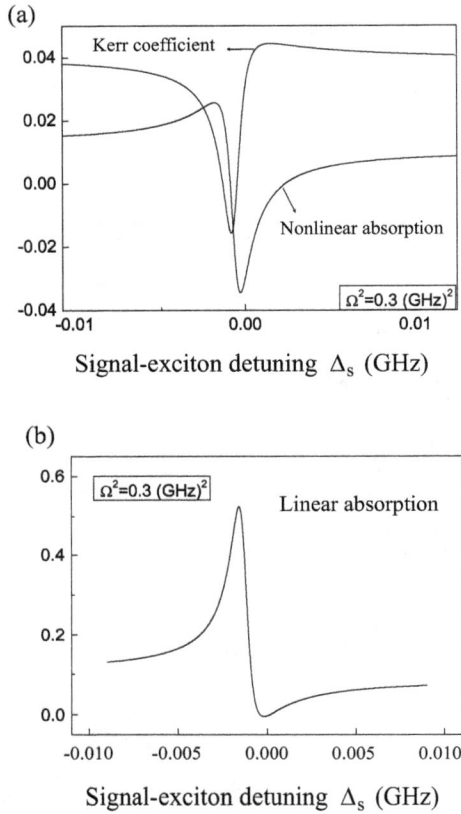

Fig. 6.14 (a) The optical Kerr coefficient and the nonlinear absorption of the signal field when $\Omega = 0.3(\text{GHz})^2$. (b) The linear absorption spectrum corresponding to the situation in (a). The parameters used are $\Delta_p = -\omega_n = -1.2\text{GHz}$, $\gamma_n = 4 \times 10^{-5}\text{GHz}$ and $\beta = 0.06$.

in Fig.6.15. We notice that this figure is similar with Fig.6.7, for the two steep peaks at the both sides correspond to the vibrational frequency of nanomechanical resonator. For example, if the frequency of nanomechanical resonator is $\omega_n = 1.2$ GHz, then the two sharp peaks at ± 1.2 GHz (solid curve). The dotted curve and dashed curve in Fig.6.15 correspond to the other two different frequencies of the nanomechanical resonator, where the sharp peaks also agree well with the resonator frequency. This means that if we first fix the control field detuning Δ_p and scan the signal frequency across the exciton frequency ω_{ex} in the spectrum, then we can easily obtain

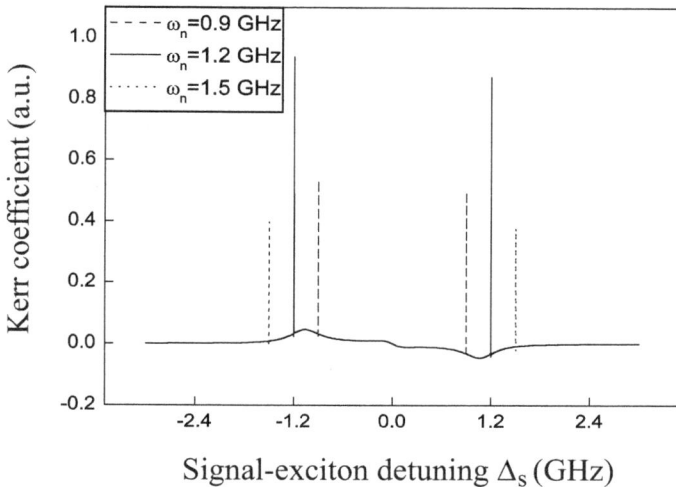

Fig. 6.15 The optical Kerr effect to measure the frequency of nanomechanical resonator as a function of the detuning Δ_s for $\Delta_p = 0$, $\gamma_n = 4 \times 10^{-5}$ GHz, $\Omega^2 = 0.3(\text{GHz})^2$ and $\beta = 0.06$.

the vibrational frequency of nanomechanical resonator exactly in nonlinear optical domain.

Bibliography

B. Lassagne, D. Garcia-Sanchez, A. Aguasca, A. Bachtold, Ultrasensitive mass sensing with a nanotube electromechanical resonator, Nano Lett. 8, 3735 (2008).

K. Jensen, K. Kim, A. Zettl, An atomic-resolution nanomechanical mass sensor, Nature Nanotech. 3, 533 (2008).

K. L. Ekinci, X. M. H. Huang, M.L. Roukes, Ultrasensitive nanoelectromechanical mass detection, Appl. Phys. Lett. 84, 4469 (2008).

D. Harris, Weighing DNA down to the zeptogram, Phys. Rev. Focus 27, 25 (2011).

J. J. Li, K. D. Zhu, Plasmon-assisted mass sensing in a hybrid nanocrystal coupled to a nanomechanical resonator, Phys. Rev. B 83, 245421 (2011).

M. D. LaHaye, O. Buu, B. Camarota, and K. C. Schwab, Approaching the quantum limit of a nanomechanical resonator, Science 304, 74 (2004).

D. Rugar, R. Budakian, H. J. Mamin and B. W. Chui, Single spin detection by magnetic resonance force microscopy, Nature 430, 329 (2004).

K. L. Ekinci and M. L. Roukes, Nanoelectromechanical systems, Rev. Sci. Instrum. 76, 061101 (2005).

I. Wilson-Rae, P. Zoller, and A. Imamoğlu, Laser cooling of a nanomechanical resonator mode to its quantum ground state, Phys. Rev. Lett. 92, 075507 (2004).

F. Jelezko, A. Volkmer, I. Popa, K. K. Rebane, and J. Wrachtrup, Coherence length of photons from a single quantum system, Phys. Rev. A 67, 041802(R) (2003).

S. D. Bennett, L. Cockins, Y. Miyahara, P. Gruetter and A. A. Clerk, Phys. Rev. Lett. 104, 017203 (2010).

T. Frey, P. J. Leek, M. Beck, A. Blais, T. Ihn, K. Ensslin and A. Wallraff, Dipole coupling of a double quantum dot to a microwave resonator, Phys. Rev. Lett. 108, 046807 (2012).

P. Q. Jin, M. Marthaler, J. H. Cole, A. Shnirman and G. Schön, Lasing and transport in a quantum-dot resonator circuit, Phys. Rev. B 84, 035322 (2011).

J. J. Li, K. D. Zhu, Mechanical vibration-induced coherent optical spectroscopy in a single quantum dot coupled to a nanomechanical resonator, J. Phys. B 43, 155504 (2010a).

J. J. Li, K. D. Zhu, A scheme for measuring vibrational frequency and coupling strength in a coupled nanomechanical resonator-quantum dot system, Appl. Phys. Lett. 94, 063116 (2009b).

J. J. Li, K. D. Zhu, An efficient optical knob from slow light to fast light in a coupled nanomechanical resonator-quantum dot system, Opt. Express 17, 19874 (2009c).

J. J. Li, K. D. Zhu, Quantum memory for light with a quantum dot system coupled to a nanomechanical resonator, Quantum Inf. Comput. 11, 0456-0465 (2011d).

J. J. Li, K. D. Zhu, A tunable optical Kerr switch based on a nanomechanical resonator coupled to a quantum dot, Nanotechnol. 21, 205501 (2010e).

K. J. Boller, A. Imamoğlu and S. E. Harris, Observation of electromagnetically induced transparency, Phys. Rev. Lett. 66, 2593 (1991).

S. E. Harris, Electromegnetically induced transparency, Phys. Today 50, 36 (1997).

M. O. Scully and M. S. Zubairy, Quantum optics, (Cambridge: Cambridge University Press) (1997).

E. Arimondo, Progress in Optics (ed. E. Wolf), Vol. 35, 257 (1996).

M. S. Bigelow, N. N. Lepeshkin and R. W. Boyd, Observation of ultraslow light propagation in a ruby crystal at room temperature, Phys. Rev. Lett. 90, 113903 (2003a).

M. S. Bigelow, N. N. Lepeshkin and R. W. Boyd, Superluminal and slow light propagation in a room-temperature solid, Science 301, 200 (2003b).

C. S. Yelleswarapu, S. Laoui, R. Philip and D. V. Rao, Coherent population oscillations and superluminal light in a protein complex, Opt. Express 16, 3844 (2008).

R. W. Boyd, Nonlinear optics, (Academic Press, Amsterdam), p. 325 (2008).

A. Zrenner, E. Beham, S. Stufler, E Findeis, M. Bichler, and G. Abstreiter, Coherent properties of a two-level system based on a quantum-dot photodiode, Nature (London) 418, 612 (2002).

S. Stufler, P. Ester, A. Zrenner, and M. Bichler, Quantum optical properties of a single $In_xGa_{1-x}As$ − GaAs quantum dot two-level system, Phys. Rev. B 72, 121301(R) (2005).

X. D. Xu, B. Sun, P. R. Berman, D. G. Steel, A. S. Bracker, D. Gammon, and L. J. Sham, Coherent optical spectroscopy of a strongly driven quantum dot, Science 317, 929 (2007).

X. D. Xu, B. Sun, E. D. Kim, K. Smirl, P. R. Berman and D. G. Steel, A. S. Bracker, D. Gammon, and L. J. Sham, Single charged quantum dot in a strong optical field: absorption, gain, and the ac-Stark effect, Phys. Rev. Lett. 101, 227401 (2008).

X. D. Xu, B. Sun, P. R. Berman, D. G. Steel, A. S. Bracker, D. Gammon, L. J. Sham, Coherent population trapping of an electron spin in a single negatively charged quantum dot, Nature Phys. 4, 692 (2008).

J. F. Lam, S. R. Forrest and G. L. Tangonan, Optical nonlinearities in crystalline organic multiple quantum wells, Phys. Rev. Lett. 66, 1614 (1991).

K. D. Zhu and W. S. Li, Electromagnetically induced transparency due to exciton phonon interaction in an organic quantum well, J. Phys. B: At. Mol. Opt. Phys. 34, L679 (2001).

I. Wilson-Rae, Intrinsic dissipation in nanomechanical resonators due to phonon tunneling, Phys. Rev. B 77, 245418 (2008).

F. Brennecke, S. Ritter, T. Donner, and T. Esslinger, Cavity optomechanics with a Bose-Einstein condensate, Science 322, 235 (2008).

R. Kanamoto, and P. Meystre, Optomechanics of a quantum-degenerate Fermi gas, Phys. Rev. Lett. 104, 063601 (2010).

H. J. Kimble, The quantum internet, Nature 453, 1023 (2008).

M. A. Hall, J. B. Altepeter, and P. Kumar, Ultrafast switching of photonic entanglement, Phys. Rev. Lett. 106, 053901 (2011).

I. C. Hoi, C. M. Wilson, G. Johansson, T. Palomaki, B. Peropadre, and P. Delsing, Demonstration of a single-photon router in the microwave regime, Phys. Rev. Lett. 107, 073601 (2011).

G. S. Agarwal, and S. Huang, Optomechanical systems as single photon routers, Phys. Rev. A 85, 021801(R) (2012).

F. L. Semião, K. Furuya, and G. J. Milburn, Nonlinear dynamics of nanomechanical and micromechanical resonators, Phys. Rev. A 79, 063811 (2009).

H. Hashemi, A. W. Rodriguez, J. D. Joannopoulos, M. Soljačić, and S. G. Johnson, Nonlinear harmonic generation and devices in doubly resonant Kerr cavities, Phys. Rev. A 79, 013812 (2009).

R. Lifshitz, and M. C. Cross, Review of Nonlinear dynamics and complexity, (Wiley) 1, 52 (2008).

H. J. R. Westra, M. Poot, H. S. J. van der Zant, and W. J. Venstra, Nonlinear modal interactions in clamped-clamped mechanical resonators, Phys. Rev. Lett. 105, 117205 (2010).

E. Hendry, P. J. Hale, J. Moger, A. K. Savchenko, and S. A. Mikhailov, Coherent nonlinear optical response of graphene, Phys. Rev. Lett. 105, 097401 (2010).

P. J. Yao, P. K. Pathak, E. Illes, S. Hughes, S. Münch, S. Reitzenstein, P. Franeck, A. Löffler, T. Heindel, S. Höfling, L. Worschech, and A. Forchel, Nonlinear photoluminescence spectra from a quantum-dot-cavity system: Interplay of pump-induced stimulated emission and anharmonic cavity QED, Phys. Rev. B 81, 033309 (2010).

P. Weinberger, John Kerr and his effects found in 1877 and 1878, Phil. Mag. Lett. 88, 897 (2008).

R. Righini, Ultrafast optical Kerr effect in liquids and solids, Science 262, 1386 (1993).

S. M. Saltiel, A. A. Sukhorukov, Y. S. Kivshar, Multistep parametric processes in nonlinear optics, Prog. Optics 47, 1 (2005).

M.-J. Li, S. Li, D. A. Nolan, Nonlinear fibers for signal processing using optical Kerr effects, J. Lightwave Technol. 23, 3606 (2005).

Z.-J. Zhong, Y. Xu, S. Lan, Q.-F. Dai, L.-J. Wu, Sharp and asymmetric transmission response in metal-dielectric-metal plasmonic waveguides containing Kerr nonlinear media, Opt. Express 18, 79 (2010).

Y. W. Jiang and K. D. Zhu, Local field effects on phonon-induced transparency in quantum dots embedded in a semiconductor medium, Appl. Phys. B 90, 79 (2008).

A. N. Vamivakas, Y. Zhao, C. Y. Lu, and M. Atature, Spin-resovled quantum-dot resonance fluorecence, Nature Phys. 5, 198 (2009).

Chapter 7

Nanomechanical Resonator Coupled to a Hybrid Nanostructure

When the sign of the real part of the dielectric function is changing across the interface of any two materials, the surface plasmons (SPs) can be produced between the interface due to the coherent electron oscillations (e.g., a metal-dielectric interface, such as a metal sheet in air) [Mares (2008); Sun (2009); Okamoto (2004)]. Applying an optical field to metallic nanostructures, the surface plasmons can provide numerous ways to manipulate light at nanoscale dimensions. Just since a decade ago, some scientific researches have shown that if coating the metallic nanoparticles (MNP) in the vicinity of quantum dot (QD) or quantum well (QW) materials, the surface plasmon produced by metallic nanoparticles after applying the external excitation, will enhance the emitting efficiency of QD/QW [Artuso (2008); Savasta (2010); Ridolfo (2011)]. Therefore, the optical devices based on semiconductor quantum dot/quantum well with coating the metallic nanoparticles will exhibit the surface plasmon enhanced emission efficiency (EE) [Gu (2011); Wang (2011); Zhmakin (2011)].

In 2006, Zhang *et al.* [Zhang (2006)] theoretically studied the optical properties of strongly coupled semiconductor and metal nanoparticle system. They predicted the existence of nonlinear Fano effect in such hybrid nanostructure system. Due to the presence of the MNP, the decay rate of the exciton will be increased. Zhang *et al.* [Zhang (2006)] showed that the plasmon-exciton interaction leads to the formation of a hybrid exciton with the shifted exciton frequency and the decreased lifetime. Afterwards, in 2008 [Yan (2008)], they investigated the hybrid semiconductor quantum dot, dye, and metal nanoparticles, which demonstrated that the multipole effects played an important role in the strong interaction regime. Recently [Zhang (2011)], Zhang *et al.* developed a quantum theory of the field-tunable nonlinear Fano effect in the hybrid metal-semiconductor nanostructures, which showed that the quantum interference due to the plasmon-exciton interaction leads to the nonlinear Fano effect. Furthermore, in a nanowire coupled to quantum dot system, Lukin's group showed a cavity-free, broadband approach for engineering photon-emitter interactions, through subwavelength confinement of optical fields near metallic nanostructures [Akimov (2007)].

Actually, there are still some interesting optical effects in the hybrid nanostructure system. In this section, we investigate a hybrid nanocrystal complex consisting of a metal nanoparticle (MNP) and a semiconductor quantum dot (SQD) embedded in a nanomechanical resonator, in the simultaneous presence of a strong pump field and a weak probe field. Compared with our previous studies in chapter 6, the resonance absorption peak and amplification peak of the probe field shown in Fig.6.3(a) are dramatically enhanced, while coupling with the metallic nanoparticle. And then the sensitivity of quantum optical devices based on this hybrid system is obviously increased. We demonstrate that the enhanced resonant absorption peaks can be continuously adjusted by the separation between the metal nanoparticle and the quantum dot, which may lead to a potential application in the technique of nanoscale optical devices such as tunable Raman lasers and bio-sensors.

7.1 Theory

In the presence of a strong pump field and a weak probe field, the hybrid complex under consideration consists of a metallic nanoparticle (MNP) and a semiconductor quantum dot (SQD) embedded in a doubly clamped suspended nanomechanical resonator as shown in Fig.7.1. As discussed in Chapter 6, at low temperatures, the SQD consists of a ground state $|g\rangle$ and the first excited state (single exciton) $|ex\rangle$ [Zrenner (2002); Stufler (2005)], which can be characterized by the pseudospin operators $S^{\pm}$ and S^z. A single gold MNP is placed above the SQD which has the radius a_0 and a center-to-center distance R from the SQD. For the nanomechanical resonator, the thickness is smaller than its width, which makes the resonator mode act as a single phonon mode with annihilation operator a and creation operator a^{+} [Wilson-Rae (2004)].

Considering that the exciton in SQD interacts with a strong pump field E_{pu} (with frequency ω_{pu}) and a weak probe field E_{pr} (with frequency ω_{pr}) simultaneously, the Hamiltonian of the hybrid system reads as follows

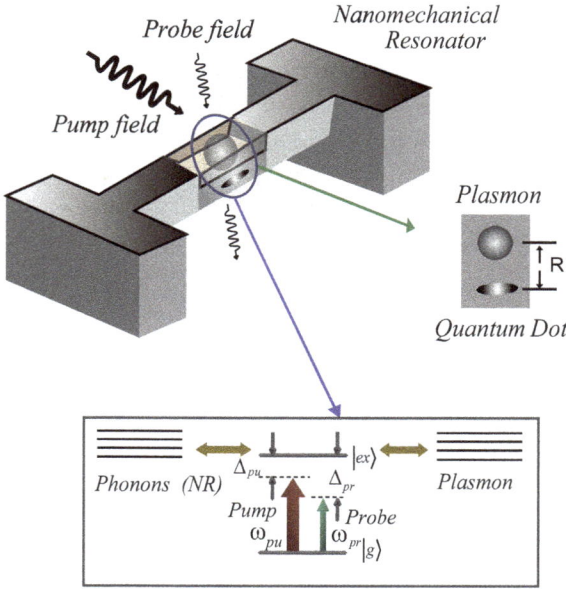

Fig. 7.1 Schematic diagram of a metal nanoparticle and a quantum dot embedded in the center of a nanomechanical resonator. The distance between MNP and QD is R. The energy-level diagram of exciton while dressing with plasmon modes and phonon modes is shown at the bottom part.

[Zhang (2006); Wilson-Rae (2004); Li (2011)]

$$H = \hbar\omega_{\mathrm{ex}}S^z + \hbar\omega_n a^\dagger a + \hbar\omega_n \beta S^z(a^\dagger + a) - \mu(E_{SQD}S^\dagger + E^*_{SQD}S^-), \quad (7.1)$$

where ω_{ex} and ω_n are the frequency of exciton and resonator mode, respectively, β is the coupling strength of the resonator mode and SQD, μ is the interband dipole matrix element and E_{SQD} is the total optical field felt by the SQD. In a rotating frame at the pump field frequency ω_{pu}, the total Hamiltonian is given by

$$H = \hbar\Delta_{\mathrm{pu}}S^z + \hbar\omega_n a^\dagger a + \hbar\omega_n \beta S^z(a^\dagger + a) - \mu(\tilde{E}_{SQD}S^\dagger + \tilde{E}^*_{SQD}S^-), \quad (7.2)$$

where $\Delta_{\mathrm{pu}} = \omega_{\mathrm{ex}} - \omega_{\mathrm{pu}}$ is the frequency difference between the exciton and pump field.

$$\tilde{E}_{SQD} = E_{\mathrm{pu}} + E_{\mathrm{pr}}e^{-i\delta t} + \frac{S_\alpha P_{MNP}}{\varepsilon_{\mathrm{eff1}}R^3}, \quad (7.3)$$

$$\varepsilon_{\mathrm{eff1}} = \frac{2\varepsilon_0 + \varepsilon_s}{3\varepsilon_0}, \quad (7.4)$$

where ε_0 and ε_s are the dielectric constants of the background medium and SQD, respectively. $\delta = \omega_{\text{pu}} - \omega_{\text{pr}}$ is the detuning of the probe and pump field. S_α is the polar factor for electric field polarization and $S_\alpha = 2$ corresponds that the polar direction is along the z axis of the hybrid system. P_{MNP} is the dipole which comes from the charge induced by the probe field. For a spherical particle whose radius is much smaller than the wavelength of light, the electric field is uniform across the particle and the electrostatic(Rayleigh) approximation is a good one. Then we have [Zhang (2006); Yan (2008)],

$$P_{MNP} = \gamma a^3 [E_{\text{pu}} + E_{\text{pr}}e^{-i\delta t} + \frac{S_\alpha P_{SQD}}{\varepsilon_{\text{eff2}} R^3}], \qquad (7.5)$$

$$\gamma = \frac{\varepsilon_{\text{Au}}(\omega) - \varepsilon_0}{2\varepsilon_0 + \varepsilon_{\text{Au}}(\omega)}, \qquad (7.6)$$

$$\varepsilon_{\text{eff2}} = \frac{2\varepsilon_0 + \varepsilon_{\text{Au}}(\omega)}{3\varepsilon_0}, \qquad (7.7)$$

$$\varepsilon_{\text{Au}}(\omega) = 1 - \frac{\omega_{\text{sp}}^2}{\omega(\omega + i\gamma_{\text{Au}})}, \qquad (7.8)$$

where ε_{Au} is the MNP's dielectric constant, ω_{sp} and γ_{Au} are the bulk metal plasma frequency and the frequency-dependent damping, respectively. The imaginary part of relative permittivity ε_{Au} determines the metallic losses [Pelton (2008); Khoo (2006)]. The dipole moment of the SQD is expressed via the off-diagonal elements of the density matrix: $P_{SQD} = \mu S^-$. The dipole approximation used here is reasonable when the distance R is large and the exciton-plasmon interaction is relatively weak [Yan (2008)]. Therefore the total optical field felt by the SQD is

$$E_{SQD} = A(E_{\text{pu}} + E_{\text{pr}}e^{-i\delta t}) + \mu B S^-, \qquad (7.9)$$

$$A = 1 + \frac{\gamma a^3 S_\alpha}{\varepsilon_{\text{eff1}} R^3}, \qquad (7.10)$$

$$B = \frac{\gamma a^3 S_\alpha^2}{\varepsilon_{\text{eff1}} \varepsilon_{\text{eff2}} R^6}. \qquad (7.11)$$

Here we use Heisenberg equation of motion $dO/dt = -i[O, H]/\hbar$, and the commutation relation $[S^z, S^\pm] = \pm S^\pm$, $[S^\dagger, S^-] = 2S^z$ and $[a, a^\dagger] = 1$. After setting $Q = a^\dagger + a$, and ignoring the quantum properties of S^z, S^-

and Q, the semiclassical equations will read as follows

$$\frac{dS^z}{dt} = -\Gamma_1(S^z + \frac{1}{2}) + i\Omega(A^*S^- - AS^\dagger) + \frac{i\mu^2 S^\dagger S^-}{\hbar}(B - B^*)$$
$$+ \frac{i\mu}{\hbar}(A^* E_{\text{pr}}^* S^- e^{i\delta t} - A E_{\text{pr}} S^+ e^{-i\delta t}), \tag{7.12}$$

$$\frac{dS^-}{dt} = -[\Gamma_2 + i(\Delta_{\text{pu}} + \omega_n \beta Q)]S^- - i2\Omega A S^z - \frac{i2\mu}{\hbar} A E_{\text{pr}} e^{-i\delta t}$$
$$- \frac{i2\mu^2 S^z S^-}{\hbar}, \tag{7.13}$$

$$\frac{d^2 Q}{dt^2} + \gamma_n \frac{dQ}{dt} + \omega_n^2 Q = -2\beta \omega_n^2 S^z, \tag{7.14}$$

where $\Omega = \mu E_{\text{pu}}/\hbar$ is the Rabi frequency of the pump field, Γ_1 and Γ_2 are the exciton relaxation rate and the exciton dephasing rate, respectively. γ_n is the decay rate of the nanomechanical resonator due to the coupling to a reservoir of "background" mode and other intrinsic processes [Ekinci (2005); Wilson-Rae (2004)]. As usual, in order to solve these equations, we make the following ansatz [Boyd (2008)]:

$$S^z = S_0^z + S_+^z e^{-i\delta t} + S_-^z e^{i\delta t}, \tag{7.15}$$

$$S^- = S_0 + S_+ e^{-i\delta t} + S_- e^{i\delta t}, \tag{7.16}$$

$$Q^- = Q_0 + Q_+ e^{-i\delta t} + Q_- e^{i\delta t}. \tag{7.17}$$

Upon substituting these equations to Eqs.(7.12)-(7.14), we can obtain the steady state equations and finally S_+, which is related to the linear optical susceptibility as $\chi_{\text{eff}}^{(1)}(\omega_{\text{pr}}) = N\mu S_+/E_{\text{pr}} = N\mu^2/\Gamma_2 \hbar \chi(\omega_{\text{pr}})$, where the dimensionless susceptibility is given by

$$\chi(\omega_{\text{pr}}) = \frac{A^2 D w_0[(2E + 2\Omega_0^2 A)(C + \delta_0) - 2B_0 \Omega_0^2 A w_0] - A w_0 G}{CG - D[A^*(D - \delta_0) + iB_{I0} A w_0][(2E + 2\Omega_0^2 A)(C + \delta_0) - 2B_0 \Omega_0^2 A w_0]}, \tag{7.18}$$

and

$$C = \Delta_{\text{pu}0} - \delta_0 - \omega_{n0}\beta^2 w_0 + B_{R0} w_0 - i(1 - B_{I0} w_0), \tag{7.19}$$

$$D = \Delta_{\text{pu}0} + \delta_0 - \omega_{n0}\beta^2 w_0 + B_{R0} w_0 + i(1 - B_{I0} w_0), \tag{7.20}$$

$$E = \Omega_0^2 \omega_{n0}\beta^2 w_0 \eta/(\Delta_{\text{pu}0} - \omega_{n0}\beta^2 w_0 + B_{R0} w_0 - i(1 - B_{I0} w_0)), \tag{7.21}$$

$$F = \Omega_0^2 \omega_{n0}\beta^2 w_0 \eta/(\Delta_{\text{pu}0} - \omega_{n0}\beta^2 w_0 + B_{R0} w_0 + i(1 - B_{I0} w_0)), \tag{7.22}$$

$$G = [A(C + \delta_0) - i2B_{I0}Aw_0][(2F + 2\Omega_0^2 A)(D - \delta_0)$$
$$- 2B_0^*\Omega_0^2 Aw_0] - iD(\Gamma_{10} - i\delta_0)(C + \delta_0)(D - \delta_0), \tag{7.23}$$

$$\eta = \frac{\omega_n^2}{\omega_n^2 - \delta^2 - i\gamma_n\delta}, \tag{7.24}$$

$$w_0 = 2S_0^z, \tag{7.25}$$

$$\delta_0 = \delta/\Gamma_2, \tag{7.26}$$

$$\Omega_0 = \Omega/\Gamma_2, \tag{7.27}$$

$$\omega_{n0} = \omega_n/\Gamma_2, \tag{7.28}$$

$$\Delta_{pu0} = \Delta_{pu}/\Gamma_2, \tag{7.29}$$

$$\Gamma_{10} = \Gamma_1/\Gamma_2, \tag{7.30}$$

$$B_0 = \mu^2 B/(\hbar\Gamma_2), \tag{7.31}$$

$$B_{R0} = \mathrm{Re}(B_0), \tag{7.32}$$

$$B_{I0} = \mathrm{Im}(B_0). \tag{7.33}$$

In the same way, the nonlinear optical susceptibility of the hybrid NR-MNP-SQD can be obtained by $\chi_{\mathrm{eff}}^{(3)}(\omega_{pr}) = (N\mu S_{-1})/(3E_{pu}^2 E_{pr}) = (N\mu^4\chi^{(3)}(\omega_{pr})/(3\hbar^3\Gamma_2^3) = \Sigma_3\chi^{(3)}(\omega_{pr})$ $(\Sigma_3 = N\mu^4/3\hbar^3\Gamma_2^3)$, where N is the number density of hybrid nanostructure, and the dimensionless nonlinear susceptibility is given as

$$\chi^{(3)}(\omega_{pr}) = \frac{(c_2 + \delta_0)|A|^2 w_0(d_2 - \delta_0 - i2B_{I0}w_0) - c_2(c_2 + \delta_0)|A|^2 w_0}{(d_2 - \delta_0)[d_2 g_2 - c_2(iA^*c_2 - iA^*\delta_0 + i2B_{I0}Aw_0)]}, \tag{7.34}$$

and

$$c_2 = \Delta_0 - \delta_0 - \omega_{n0}\beta^2 w_0 + B_{R0}w_0 + i(1 - B_{I0}w_0), \tag{7.35}$$

$$d_2 = \Delta_0 + \delta_0 - \omega_{n0}\beta^2 w_0 + B_{R0}w_0 - i(1 + B_{I0}w_0), \tag{7.36}$$

$$e_2 = \Omega_0^2\omega_{n0}\beta^2 w_0\eta^*/(\Delta_0 - \omega_{n0}\beta^2 w_0 + B_{R0}w_0 - i(1 + B_{I0}w_0)), \tag{7.37}$$

$$f_2 = \Omega_0^2\omega_{n0}\beta^2 w_0\eta^*/(\Delta_0 - \omega_{n0}\beta^2 w_0 + B_{R0}w_0 + i(1 - B_{I0}w_0)), \tag{7.38}$$

$$g_2 = -\frac{ic_2(\Gamma_{10} + i\delta_0)(c_2 + \delta_0)(d_2 - \delta_0)}{[(d_2 - \delta_0)(2Ae_2 + 2\Omega_0^2 A) - 2B_0\Omega_0^2 Aw_0]}$$
$$+ \frac{[(2Af_2 + 2\Omega_0^2 A^*)(c_2 + \delta_0) - 2B^*\Omega_0^2 Aw_0](Ad_2 - A\delta_0 - i2B_{I0}Aw_0)}{[(d_2 - \delta_0)(2Ae_2 + 2\Omega_0^2 A) - 2B_0\Omega_0^2 Aw_0]}. \tag{7.39}$$

The population inversion of exciton(w_0) is determined by the equation

$$(w_0+1)[(1-B_{I0}w_0)^2+(\Delta_{\text{pu}0}-\omega_{n0}\beta^2 w_0+B_{R0}w_0)^2]+2\Omega_0^2 A^2 w_0 = 0. \quad (7.40)$$

7.2   Coherent optical spectrum enhancement

We choose the realistic coupled system of an InAs quantum dot, gold MNP and a GaAs nanomechanical resonator. In this case $\Gamma_1 = 2\Gamma_2 = 0.3$GHz, $\omega_n = 1.2$GHz, $\gamma_n = 1 \times 10^{-4}$GHz, $\beta = 0.06$ [Wilson-Rae (2004)], $a_0 = 2.5$nm, $\mu = 40D$, $\varepsilon_0 = 1$, $\varepsilon_s = 6$, $\omega_{\text{sp}} = 1.37 \times 10^3$THz and $\gamma_{\text{Au}} = \omega_{\text{sp}}/60$ [Zhang (2006); Yan (2008); Pelton (2008); Khoo (2006)]. Figure 7.2 shows the probe absorption spectrum as a function of the detuning Δ_{pr} between probe field and exciton for several separations.

The solid curve in Fig.7.2(a) shows that when the MNP is placed far away from the SQD ($R \to \infty$), there is a more separated absorption peak, which is the same with Fig.6.5. But when approaching the MNP to the SQD, the peak splitting will shrink due to the coupling between MNP and SQD. As the probe field frequency is close to the resonance frequency of gold MNP, the probe field will excite the collective surface charge oscillations on MNP. The MNP acts as an antenna and radiates the exponential spatial distribution field. The enhanced field will promote the absorption of the excitation light by the SQD, and make the SQD much easier to saturate which causes the peak splitting become narrower. Figure 7.2(b) shows the relationship between the peak splitting shown in Fig.(a) and the separation between SQD and MNP. It is obvious that the splitting has a nearly linear dependence on the separation between MNP and SQD, which can be used to describe the coupling strength of the hybrid system. In this case, Fig.7.2 provides us a method to measure the coupling strength between MNP and SQD according to the peak splitting in probe absorption spectrum.

Figure 7.3(b) shows the probe absorption spectrum as a function of Δ_{pr} in the case $\Delta_{\text{pu}} = 0$, without the interaction between MNP and SQD. It is clear that the two sharp peaks appearing at sidebands of probe absorption spectrum just exactly locate at $\delta = \pm\omega_n$, which is the same with Fig.6.3(a). However, Fig.7.3(a) and (c) show that the absorption peak and amplification peak will be dramatically enhanced when SQD couples to the MNP, due to the plasmon enhancement effect. The plasmon polarizes the exciton in SQD, and increases the population inversion of exciton. Then the absorption of the light by SQD and radiative emission rates will increase. When the exciton decays finally, it will emit more intensive fields than the

(a)

(b)

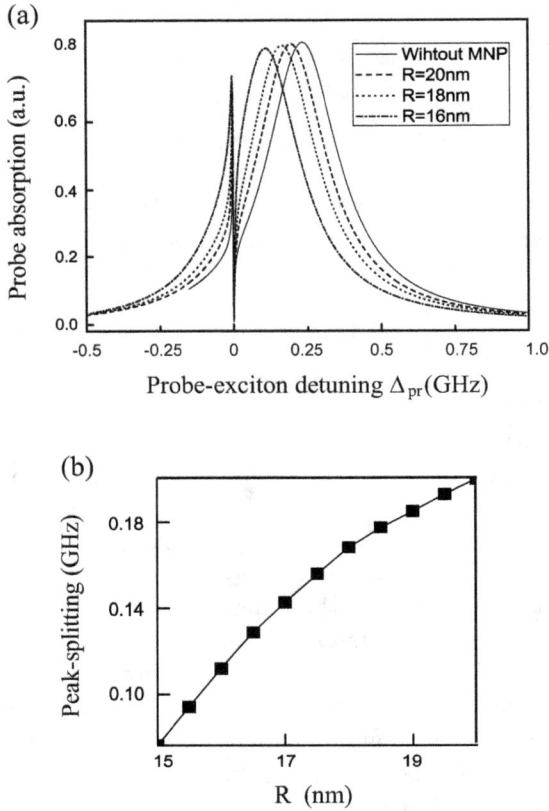

Fig. 7.2   (a) The probe absorption spectrum as a function of the detuning Δ_{pr} for several separations. (b) The relationship between the peak splitting and the center-to-center distance R between MNP and SQD. The parameters are $\Gamma_1 = 2\Gamma_2 = 0.3\mathrm{GHz}$, $\omega_n = 1.2\mathrm{GHz}$, $\gamma_n = 1 \times 10^{-4}\mathrm{GHz}$, $\beta = 0.06$, $\Delta_{\mathrm{pu}} = 1.2\mathrm{GHz}$, $\Omega^2 = 0.09(\mathrm{GHz})^2$, $a_0 = 2.5\mathrm{nm}$, $\mu = 40D$, $\omega_{\mathrm{sp}} = 1.37 \times 10^3\mathrm{THz}$ and $\gamma_{\mathrm{Au}} = \omega_{\mathrm{sp}}/60$.

system without MNP. As shown in Fig.7.3(a) and (c), the enhancement can be continuously adjusted by the separation of the MNP and SQD. When the separation is 16 nm, the enhanced peak will be almost 3 times larger than the system without MNP. From Fig.7.3(a) and (c), we can also find that the full width at half maximum of the peak will reduces as the separation diminishes.

The amplification enhancement and absorption enhancement as a function of separation between SQD and gold MNP are shown in Fig.7.4.

We notice that, as the separation between MNP and SQD deceases, the gain enhancement has an exponential increase. Especially, at a separation of 14.8nm, about three orders of magnitude of the enhancement in

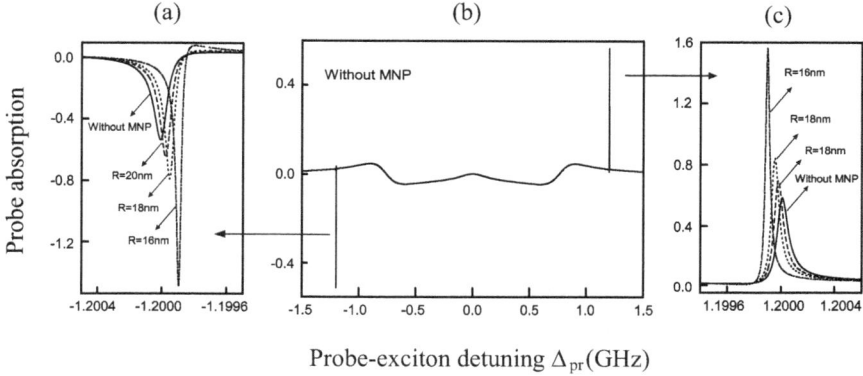

Fig. 7.3 (a) The enlarged view of negative absorption peaks of probe field, for different separations between MNP and SQD. (b) The absorption spectrum of the probe field as a function of probe-exciton detuning without the interaction between MNP and SQD. (c) The resonance absorption peaks shown in (b), for different separations between MNP and SQD. The parameters used are $\Gamma_1 = 2\Gamma_2 = 0.3$GHz, $\omega_n = 1.2$GHz, $\gamma_n = 1 \times 10^{-4}$GHz, $\beta = 0.06$, $\Delta_{pu} = 0$ and $\Omega^2 = 0.09$(GHz)^2, $a_0 = 2.5$nm, $\mu = 40D$, $\omega_{sp} = 1.37 \times 10^3$THz and $\gamma_{Au} = \omega_{sp}/60$.

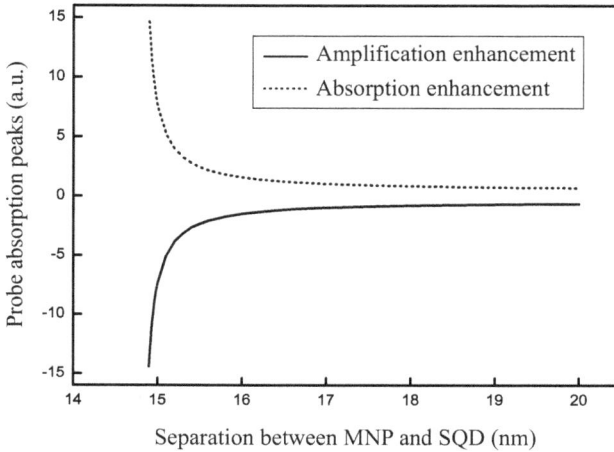

Fig. 7.4 The amplification enhancement and absorption enhancement as a function of the separation between gold MNP and SQD. Other parameters are the same as in Fig.7.3.

amplification peak can be achieved in this coupled system. The tunable amplification enhancement proposed here may lead to a potential application in Raman lasers.

7.3   All-optical Kerr modulator

As discussed in Chapter 6, the nonlinear optical effect is the hot pursuit in quantum optics. Here we demonstrate that the nonlinear optical Kerr effect based on the coupled NR-MNP-SQD system is obviously enhanced compared with NR-SQD interaction in Chapter 6. The plasmon resonance of an Au nanoparticle will enhance the nonlinear interaction between a quantum dot and a nanomechanical resonator, resulting in several orders of larger enhancement in optical Kerr effect than intrinsic silicon and many other synthesized structures. Furthermore, the optical Kerr nonlinearity can be tuned simply by the intensity of pump field. Such an advantage may have the potential applications in ultrafast optical Kerr modulator.

We use the same parameters as in Fig.7.2. Figure 7.5 shows the linear absorption, the nonlinear absorption, and the optical Kerr nonlinearity

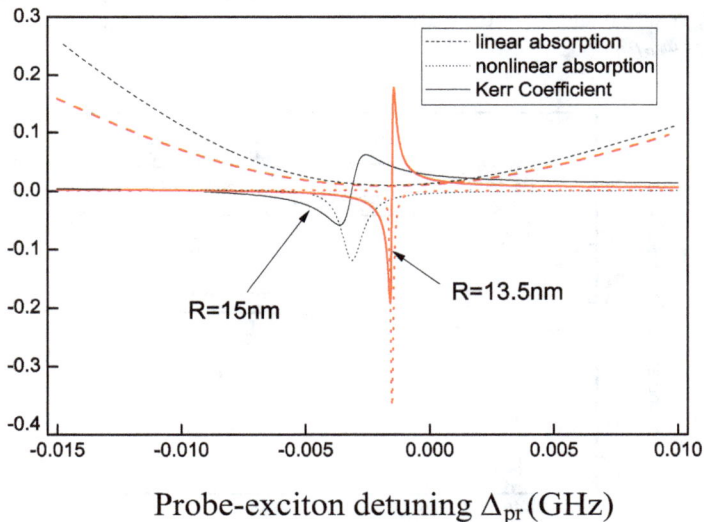

Fig. 7.5   Plots of the linear absorption, nonlinear absorption and optical Kerr nonlinearity as a function of probe-exciton detuning in two different center-to-center distances between MNP and SQD ($R = 15$nm and $R = 13.5$nm).

as a function of probe-exciton detuning in two different center-to-center distances between MNP and SQD. It is seen that under the working region of the Kerr nonlinearity, the nonlinear absorption is always negative in two different separations of MNP and SQD. The properties ensure the accessibility of large Kerr effect with little absorption occurring in experiment. Actually, the former works of Wang and Lv [Wang (2005); Lv (2008)] have shown that the surface plasmon resonance does affect the linear and nonlinear absorption, and the enhancement of third-order nonlinear optical properties in metal nanocomposites is observed in experiment.

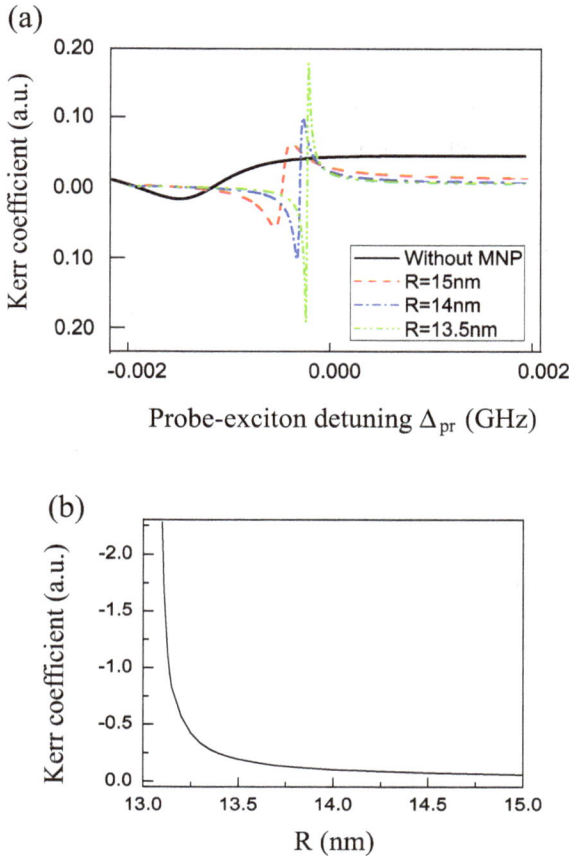

Fig. 7.6 (a) The optical Kerr coefficient as a function of probe-exciton detuning without and with MNP. (b) The relationship between Kerr coefficient and the MNP-NR distance *R*.

Figure 7.6 demonstrates the optical Kerr nonlinearity as a function of probe-exciton detuning with or without MNP. When the MNP is placed far away from the SQD, the optical Kerr effect produced in Fig.7.6 shows the coupling between SQD and nanomechanical resonator, which was discussed in Fig.6.13. However, as the MNP approaches the SQD, the probe field excites the collective surface plasmon on MNP, which will enhance the nonlinear interaction between SQD and leads to the significant optical Kerr coefficient. The MNP plays a role as an antenna and radiates the exponential spatial distribution field, therefore the enhancement will be more obvious as the separation between SQD and MNP decreases. Figure 7.6(b) indicates the relationship between Kerr coefficient and the center-to-center distance R, which shows that the Kerr coefficient has an exponential enlargement as long as the MNP approaches the SQD. Figure 7.6(b) shows that the Kerr coefficient has a ten-fold enhancement at $R = 13.5$nm, even more, a 100-fold enhancement at $R = 13$nm.

Here, the value of optical Kerr coefficient can be estimated as $\mathrm{Re}\chi_{\mathrm{eff}}^{(3)}(\omega_s) = 4.1 \times 10^{-15}\mathrm{m}^2/\mathrm{V}^2$, due to $N = 10^9\mathrm{m}^{-3}$ [Li (2009)], and $\mu = 28D$ [Vamivakas (2009)]. The optical Kerr coefficient in this nanocrystal complex coupled to a nanomechanical resonator compared with ones in other systems is given in Table 1. It is obvious that the optical Kerr coefficient value in the hybrid MNP-SQD-NR system is several orders of magnitude larger than those in the other systems.

Figure 7.7 shows the optical Kerr coefficient as a function of the pump field intensity Ω^2 with two different center-to-center distances. We notice that when fixing the separation between SQD and MNP at $R = 13$nm, the optical Kerr coefficient will increase slowly at first, and then increase dramatically after the threshold $0.27(\mathrm{GHz})^2$. The same situation occurs in $R = 14$nm, while the threshold presents at $0.17(\mathrm{GHz})^2$. Figure 7.7 shows that the optical Kerr nonlinearity can be modulated effectively by the intensity of pump field.

Table 1.   Different optical Kerr coefficients with different monocrystal complexes.

Work	System	optical Kerr coefficient
Ref.[Dinu (2003)]	Intrinsic Silicon	$4.5 \times 10^{-18}\mathrm{m}^2/\mathrm{V}^2$
Ref.[Lu (2008)]	SQD@MNP	$0.8 \times 10^{-21}\mathrm{m}^2/\mathrm{V}^2$
Ref.[Li (2009)]	SQD@NR	$1.4 \times 10^{-17}\mathrm{m}^2/\mathrm{V}^2$
Our work	SQD@NR@MNP	$4.1 \times 10^{-15}\mathrm{m}^2/\mathrm{V}^2$

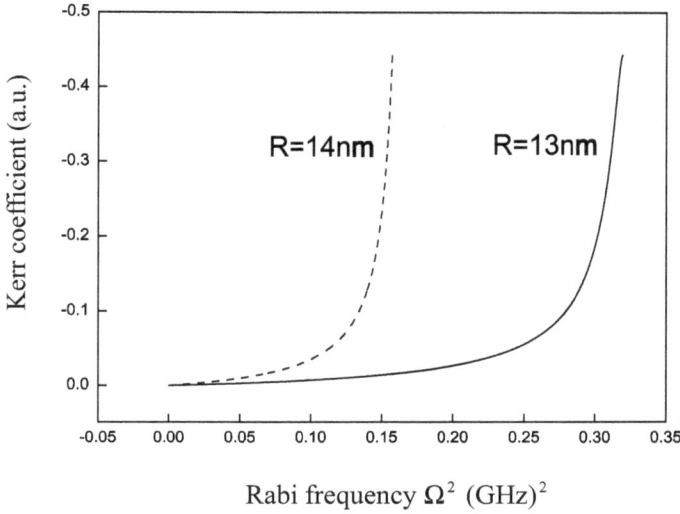

Fig. 7.7   The optical Kerr coefficient as a function of the pump field intensity Ω^2 in two different center-to-center distances R. The parameters are the same as in Fig.7.2.

7.4   Surface plasmon enhanced optical mass sensing

Taking advantages of surface plasmon for spectrum enhancement and the coupled SQD-NR system discussed in chapter 6, in this section, we propose an optical mass sensing based on a hybrid MNP-SQD embedded in a doubly clamped nanomechanical resonator, driven by two optical fields. The nanomechanical resonators act as mass sensors due to that their resonant frequency is sensitive to the mass adsorbed to them. The mass sensing monitors the shift $\Delta\omega_n$ of ω_n induced by the adsorption onto the resonator of the molecular species to weigh. $\Delta\omega_n$ is related to the deposited mass m_d by [Ekinci (2004)]

$$m_d = (2\frac{m_n}{\omega_n})\Delta\omega_n. \tag{7.41}$$

This is the direct relationship between the external accreted mass and the frequency shift of the nanomechanical resonator. Therefore, if we measure the frequency shift in the probe absorption spectrum exactly, the accreted mass can be obtained accurately. Taking [Jensen (2008); Lassagne (2008); Ekinci (2004)] advantage of the localized surface plasmon, the signal of frequency-shift is largely enhanced in the probe absorption spectrum using the pump-probe technique.

In the following, using the coupled system of the MNP-QD and the nanomechanical resonator, we illustrate how to weigh the mass of accreted particles landing on the nanomechanical resonator in an all-optical domain. Figure 7.1 shows our proposed setup of the plasmon-assisted mass sensing in the presence of a strong pump laser and a weak probe laser. The inset in Fig.7.1 shows the energy levels of the exciton when dressing with the surface plasmon and the vibration of nanomechanical resonator. In our pump-probe technique, we first aim a strong pump laser at this coupled system, and then detect the probe absorption spectrum while the probe laser is applied.

As the same way, we depict the probe absorption spectrum in Eq.(7.18) as a function of probe-exciton detuning Δ_{pr} with $\Delta_{pu} = 0$ as shown in Fig.7.8(a). From this curve we find that there are two sharp peaks at both sides of the spectrum which just correspond to the vibrational frequency of the nanomechanical resonator. Particularly, for our selected GaAs NR, the vibrational frequency is $\omega_n = 1.2$GHz, then the two sharp peaks appear at ± 1.2GHz in the probe absorption spectrum, respectively. The two sharp peaks represent the resonance amplification and absorption of the vibrational mode of the NR. The underlying physical mechanism for this phenomenon can be understood as: the simultaneous presence of the pump and probe fields generates a beat wave oscillating at the beat frequency $\delta = \omega_{pr} - \omega_{pu}$ to drive the nanomechanical resonator via the hybrid SQD-MNP. If the beat frequency δ is close to the resonance frequency ω_n of the NR, the nanomechanical resonator starts to oscillate coherently, which will result in Stokes ($\omega_s = \omega_{pu} - \omega_n$) and anti-Stokes ($\omega_{as} = \omega_{pu} + \omega_n$) scattering of light from the pump field. For the near-resonant probe laser, the probe field will interfere with the Stokes field and the anti-Stokes field, respectively. As a result, the probe spectrum can be modified significantly. Therefore, it is convenient to obtain the vibrational frequency of the nanomechanical resonator in the probe spectrum by scanning the probe frequency across the exciton frequency, provided that the pump-exciton detuning is fixed to zero.

Figure 7.8(b) shows the enlarged view of the right absorption peak and left amplification peak, with different decay rates of nanomechanical resonator, which indicates that one can narrow the spectral width of the probe spectrum and enhance the probe spectrum via decreasing the decay rate of the NR. Here, since the $\gamma_n = 4 \times 10^{-5}$GHz, the narrow spectral width ~ 100kHz provides an enough resolution to do mass sensing. In this case, the better sensitivity of mass sensing can be obtained if we select a smaller decay rate of the NR.

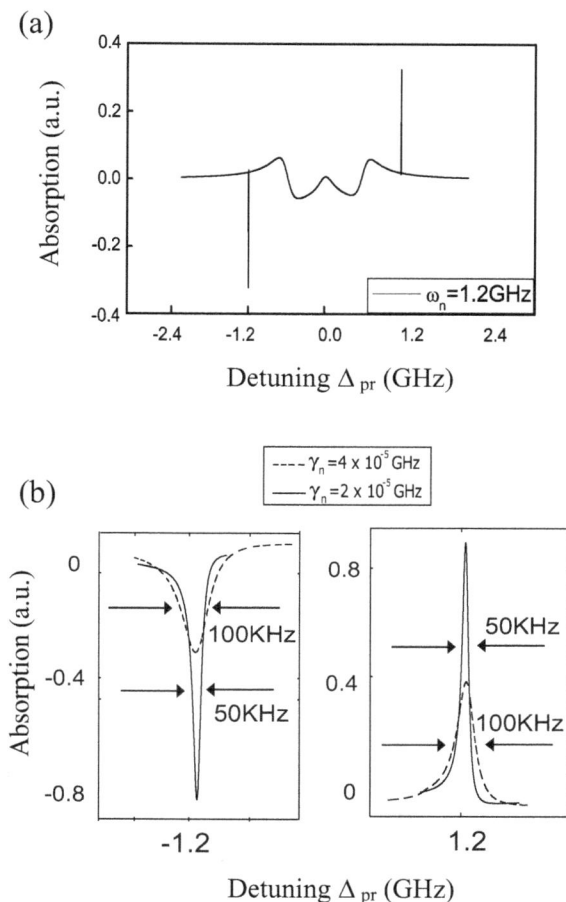

Fig. 7.8   (a) The probe absorption spectrum as a function of detuning Δ_{pr} for the case $\Omega^2 = 0.05(\mathrm{GHz})^2$ (corresponds to the pump intensity $I = 36\mathrm{mW/cm}^2$), $\omega_n = 1.2\mathrm{GHz}$, $\Delta_{\mathrm{pu}} = 0$, $\gamma_n = 4 \times 10^{-5}\mathrm{GHz}$, $\beta = 0.06$, $m_n = 5.3 \times 10^{-15}g$, $a_0 = 2.5\mathrm{nm}$, and $R = 16\mathrm{nm}$. (b) The enlarged view of the right peak and the left peak shown in (a). The dashed curve curve and the solid curve for $\gamma_n = 4 \times 10^{-5}\mathrm{GHz}$ and $\gamma_n = 2 \times 10^{-5}\mathrm{GHz}$ have significantly different spectral widths (100kHz and 50kHz).

Since we have measured the vibrational frequency of nanomechanical resonator, the next step is to determine the frequency-shift while landing the deposited nanoparticle onto the surface of NR. The physical layout of the entire plasmon-assisted mass sensing apparatus, including nanomechanical resonator, hybrid MNP-SQD, evaporation system and two optical

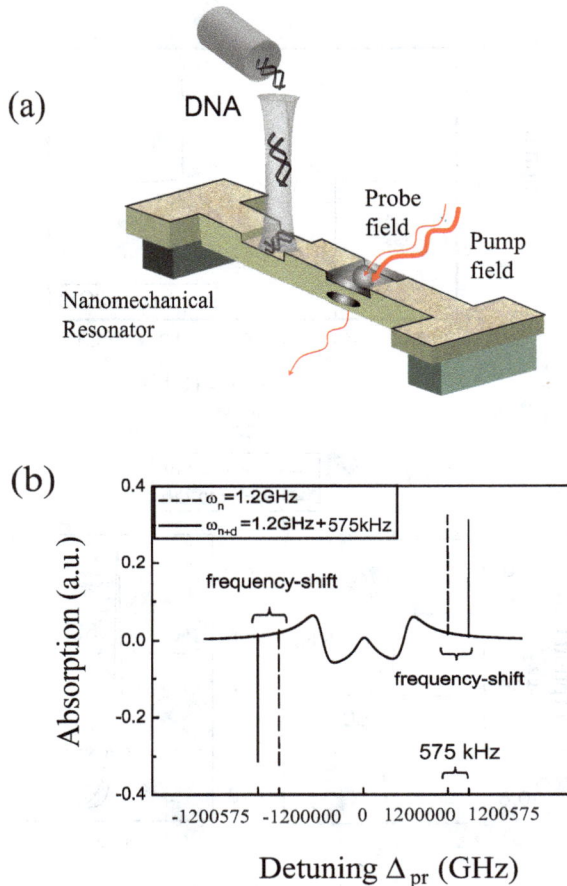

Fig. 7.9 (a) The simplified process for weighing the mass of Escherichia coli strain CA46 plasmid pColG DNA with plasmon-assisted mass sensor, driven by two optical fields. (b) The probe absorption spectrum while landing the DNA molecule on the nanomechanical resonator. The dashed curve and the solid curve are the probe absorption spectrum without and with landing DNA molecule, respectively. The parameters used are the same with Fig.7.8.

lasers, is shown in Fig.7.9(a). All the experiments are done in situ within a cryogenically cooled, ultrahigh vacuum apparatus with ultralow pressure, such as below 10^{-10} Torr.

In traditional mass spectrometry, in order to know the mass, the analysis molecules have to be ionized before their mass is extracted. But not all molecules are suitable for ionizing, such as long-chain bioactive molecules.

In this case, in order to give a guidance to experiment, we choose the landing particle as a realistic DNA molecule of the Escherichia coli strain CA46 plasmid pColG with 4715bp [Website (NCBI)]. We put this pColG DNA upon the surface of nanomechanical resonator in an apparatus depicted in Fig.7.9(a), and then repeat the first step. Assuming the mass of DNA molecule landing on the nanomechanical resonator is uniformly distributed. The two new resonant peaks will appear in the probe absorption spectrum, which have a slightly frequency shift from their original position in the first step without landing the DNA molecule. Figure 7.9(b) illustrates the physical picture of frequency change, where the dashed curve is the original probe spectrum without landing the particle, while the solid curve shows the new frequency of nanomechanical resonator after depositing pColG DNA molecules. According to [Website (NCBI)], the mass of the Escherichia coli strain plasmid is $4715 \times 2 \times 324.5/(6.022 \times 10^{23}) = 5081$zg (1 zg$=10^{-21}$g). Here we notice that the frequency-shift is 575kHz. According to Eq.(7.40), for pColG plasmid DNA with $\Delta\omega_n = m_d\omega_n/2m_n = 575$kHz, the new resonant steep peaks will appear at ±1200575kHz. The narrow spectral width in Fig.7.9(b) provides enough spectral resolution for this mass sensing. Otherwise, the mass of multiple molecules can also be detected via landing them together onto the surface of nanomechanical resonator. It is clear that the frequency-shift of the landing multiple particles is larger than that of the landing single particle.

Compared with traditional electrical methods, the main benefits of our proposed plasmon assisted optical mass sensing, driven by two optical fields, are listed as follows: (1) operating at high frequencies; (2) avoiding the heating effect or energy loss caused by circuits; (3) narrowing spectral width due to surface plasmon. Particularly, the spectral width of the probe spectrum in our technique is on the order of magnitude of kHz, while the electric methods in [Naik (2009); Chiu (2008)] gave the spectral width on the order of magnitude of MHz. In this case, the plasmon assisted mass sensing driven by two optical fields has better sensitivity than traditional electrical methods.

Bibliography

J. W. Mares, M. Falanga, A. V. Thompson, A. Osinsky, J. Q. Xie, B. Hertog, A. Dabiran, P. P. Chow, S. Karpov, W. V. Schoenfeld, Hybrid CdZnO/GaN quantum well light emitting diodes. J. Appl. Phys. 104, 093107 (2008).

Q. J. Sun, G. Subramanyam, L. M. Dai, M. Check, A. Cambell, R. Naik, J. Grote, Y. Q. Wang, Highly efficient quantum-dot light-emitting diodes with DNACTMA as a combined hole-transporting and electron-blocking layer. ACS Nano 3, 737 (2009).

K. Okamoto, I. Niki, A. Shvartser, Y. Narukawa, T. Mukai, A. Scherer, Surface-plasmon-enhanced light emitters based on InGaN quantum wells. Nat. Mater. 3, 601 (2004).

R. D. Artuso, and G. W. Bryant, Optical response of strongly coupled quantum dot-metal nanoparticle system: Double peaked Fano structure and bistability. Nano Lett. 8, 2106 (2008).

S. Savasta, R. Saija, A. Ridolfo, O. Di Stefano, P. Denti, F. Borghese, Nanopolaritons: Vacuum rabi splitting with a single quantum dot in the center of a dimer nanoantenna. ACS Nano 4, 6369 (2010).

A. Ridolfo, R. Saija, S. Savasta, P. H. Jones, M. Antonia Iatì, and O. M. Maragò, Fano-doppler laser cooling of hybrid nanostructures. ACS Nano 5, 7354 (2011).

X. Gu, T. Qiu, W. Zhang, P. K Chu, Light-emitting diodes enhanced by localized surface plasmon resonance. Nanoscale Res. Lett. 6, 199 (2011).

Z. M. Wang, A. Neogi (eds.), Nanoscale Photonics and Optoelectronics. pp. 27, Springer (2011).

A. I. Zhmakin, Enhancement of light extraction from light emitting diodes. Phys. Rep. 498, 189 (2011).

W. Zhang, A. O. Govorov, and G. W. Bryant, Semiconductor-metal nanoparticle molecules: hybrid excitons and the nonlinear Fano effect. Phys. Rev. Lett. 97, 146804 (2006).

J. Y. Yan, W. Zhang, S. Duan, and A. O. Govorov, Optical properties of coupled metal-semiconductor and metal-molecule nanocrystal complexes: Role of multipole effects. Phys. Rev. B 77, 165301 (2008).

W. Zhang, and A. O. Govorov, Quantum theory of the nonlinear Fano effect in hybrid metal-semiconductor nanostructures: The case of strong nonlinearity. Phys. Rev. B 84, 081405 (2011).

A. V. Akimov, A. Mukherjee, C. L. Yu, D. E. Chang, A. S. Zibrov, P. R. Hemmer, H. Park, M. D. Lukin, Generation of single optical plasmons in metallic nanowires coupled to quantum dots. Nature 450, 402 (2007).

A. Zrenner, E. Beham, S. Stufler, F. Findeis, M. Bichler, and G. Abstreiter, Coherent properties of a two-level system based on a quantum-dot photodiode. Nature(London) 418, 612 (2002).

S. Stufler, P. Ester, A. Zrenner, and M. Bichler, Quantum optical properties of a single $In_xGa_{1-x}As$-GaAs quantum dot two-level system. Phys. Rev. B 72, 121301 (2005).

I. Wilson-Rae, P. Zoller, and A. Imamoğlu, Laser cooling of a nanomechanical resonator mode to its quantum ground state. Phys. Rev. Lett. 92, 075507 (2004).

J. J. Li, K. D. Zhu, Plasmon-assisted mass sensing in a hybrid nanocrystal coupled to a nanomechanical resonator. Phys. Rev. B 83, 245421 (2011).

M. Pelton, J. Aizpurua and G. Bryant, Metal-nanoparticle plasmonics. Laser & Photon. Rev. 2, 136 (2008).

I. C. Khoo, D. H. Werner, X. Liang, A. Diaz and B. Weiner, Nanosphere dispersed liquid crystals for tunable negative-zero-positive index of refraction in the optical and terahertz regimes. Opt. Lett. 31, 2592 (2006).

K. L. Ekinci and M. L. Roukes, Nanoelectromechanical systems. Rev. Sci. Instrum. 76, 061101 (2005).

R. W. Boyd, Nonlinear optics. (Academic, San Diego) p. 313 (2008).

Y. Wang, X. B. Xie, G. Goodson, Enhanced third-order nonlinear optical properties in dendrimer-metal nanocomposites. Nano Lett. 5, 2379 (2005).

J. Lv, L. Jiang, C. H. Li, X. F. Liu, M. J. Yuan, J. L. Xu, W. D. Zhou, Y. L. Song, H. B. Liu, Y. L. Li, D. B. Zhu, Large Third-order optical nonlinear effects of gold nanoparticles with unusual fluorescence enhancement. Langmuir 24, 8297 (2008).

J. J. Li, K. D. Zhu, A scheme for measuring vibrational frequency and coupling strength in a coupled annomechancial resonator-quantum dto system. Appl. Phys. Lett. 94, 063116 (2009); 94, 249903 (2009).

A. N. Vamivakas, Y. Zhao, C. Y. Lu, M. Atatire, Spin-resolved quantum-dot resonance fluorescence. Nat. Phys. 5, 198 (2009).

M. Dinu, F. Quochi, H. Garcia, Third-order nonlinearities in silicon at telecom wavelengths. Appl. Phys. Lett. 82, 2954 (2003).

Z. E. Lu, and K. D. Zhu, Enhancing Kerr nonlinearity of a strong coupled exciton-plasmon in hybrid nanocrystal molecules. J. Phys. B 41, 185503 (2008).

K. L. Ekinci, Y. T. Yang and M. L. Roukes, Ultimate limits to inertial mass sensing based upon nanoelectromechanical systems. J. Appl. Phys. 95, 2682 (2004).

K. Jensen, K. Kim, A. Zettl, An atomic-resolution nanomechanical mass sensor. Nat. Nanotechnol. 3, 533 (2008).

B. Lassagne, D. Garcia-Sanchez, *et al.* Ultrasensitive mass sensing with a nanotube electromechanical resonator. Nano Lett. 8, 3735 (2008).

K. L. Ekinci, X. M. H. Huang, M. L. Roukes, Ultrasensitive nanoelectromechanical mass detection. Appl. Phys. Lett. 84, 4469 (2004).

NCBI, *http://www.ncbi.nlm.nih.gov/nuccore/NC_010904.1.*

A. K. Naik, M. S. Hanay, W. K. Hiebert, X. L. Feng and M. L. Roukes, Towards single-molecule nanomechanical mass spectrometry. Nat. Nanotechnol. 4, 445 (2009).

H. Y. Chiu, P. Hung, H. W. Ch. Postma and M. Bockrath, Atomic-scale mass sensing using carbon nanotube resonators. Nano Lett. 8, 4342 (2008).

Chapter 8

Optomechanical System with a Carbon Nanotube Resonator

Carbon nanotubes (CNTs), nearly ideal one-dimensional (1D) systems, are allotropes of carbon with a cylindrical nanostructure [Khlobystov (2011)]. These cylindrical structures made CNTs have wide range of applications in optics, electronics, engineering science and material technology [Tanaka (2011); Wu (2011); Han (2010)]. The honeycomb structure of carbon-atom, made CNTs treated as direct bandgap semiconductors, or nearly ballistic conduction metals. The structure of a semiconducting carbon nanotube can be viewed as a graphene rolled into a cylinder, with diameters of only 1-3nm and lengths that can be on the scale of centimetres [Mueller (2009); Shin (2009); Avouris (2008)]. An optical or electrical field applying on the semiconducting CNTs can produce the excited states and some excitons. These exitons, with dissociation energies of around 0.5eV, are luminescent and one-dimensional. When modulating the Stark-shift, the mass of the exciton is localized in the center of CNTs [Avouris (2008)].

Carbon nanotube resonators are particularly attractive for their unique features of strong nonlinear response, long vibration lifetime, easy fabrication and high quality factor [Avouris (2008); Muoth (2010)]. The insensitive properties to external temperature make the transmission spectrum enhanced and the environment noise reduced, while detecting the optical effects of the CNT. Consequently, CNT resonators enable ultrasensitive photonic sensors as a possible replacement for silicon-based transistors, such as photovoltaics, nanotherapeutics, bioimaging, and superconducting devices [Singhal (2010); Belzig (2010); Zhang (2011); Farhat (2011)]. The most interesting feature for seminconducting nanotubes is the generating and detection of light, which provides the potential of some unified electronic and optoelectronic CNT devices, such as transistors, light controllers or light amplifiers, etc.

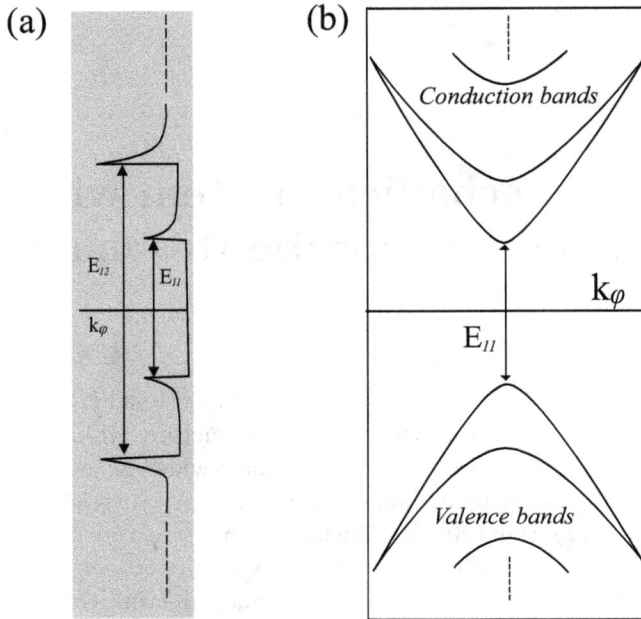

Fig. 8.1   (a) The density states and (b) bound states of single-particle model of carbon nanotube. k_φ is the wave vector along the CNT axis. E_{11} and E_{22} are the first allowed transition and the second allowed transition, respectively.

In this chapter, we investigate the optical properties of a doubly clamped carbon nanotube resonator in the presence of two optical fields, as shown in Fig.8.2(b). Figure 8.1 shows the energy state of carbon nanotube where the momentum is finite. For a single-particle model in Fig.8.1, the electron-hole interactions allow formation of bound electron-hole pairs, where one or a series of bound states (so-called excitons) are formed. The exciton binding energies in nanotubes have been predicted, and are found to be large, as great as several hundred meV. However, in the absence of a magnetic field there is one single bright level: the singlet E_{11} direct exciton [Ando (2006)]. Therefore, the carbon nanotube acts like a quantum dot when electrons are confined to a small region of carbon nanotube [Yasukochi (2011)]. We analyze a tip electrode configuration that effectively engineers a pair of tunable optically active nanotube quantum dots (NTQDs) with excitonic level spacing in the meV range corresponding to a confinement length below 10nm. On one hand, the quantum confinement is induced by the inhomogeneity in the field component along the CNT axis $E_{//}$, as shown in Fig.8.2(b). On the other hand, the normal component $E_\perp$ can be used to induce a tunable parametric coupling between the localized exciton and the flexural motion of the CNT [Wilson-Rae (2009)]. This localized exciton is formed in the segment of nanotube between the doubly clamped suspensions, leading to a quantized energy spectrum in the longitudinal direction.

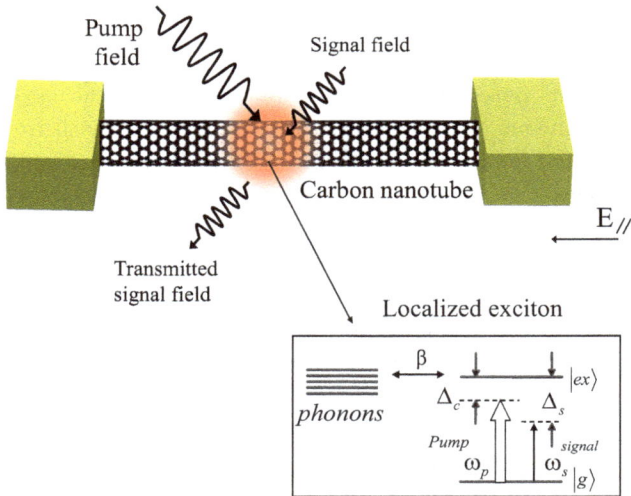

Fig. 8.2 Comparison of (a) cavity optomechanical system and (b) carbon nanotube optomechanical system. The suspended carbon nanotube resonator in (b) is driven by a strong pump laser and probed by a weak signal laser. The inset describes the energy level of localized exciton while dressing with the vibration modes of carbon nanotube resonator. Here, the mechanical vibration of carbon nanotube coupling to localized exciton via deformation potential is in analogy with the mechanical oscillator coupling to optical cavity via radiation pressure in optomechanical system.

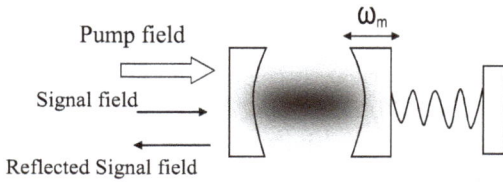

Here, for optomechanical system with a carbon nanotube in Fig.8.2(b), the role of the optical cavity is played by an excitonic resonance of the CNT that couples parametrically to the motion via deformation potential electron-phonon interactions. After studying the optical properties of a suspended carbon nanotube, we demonstrate the realization of the slow light and superliminal light in a suspended carbon nanotube, with the pump-signal technique. Detailed analysis shows that the linewidth of signal light spectrum is determined by the vibration decay rate of carbon nanotube. The smaller the decay rate of carbon nanotube is, the narrower the signal spectral linewidth is. We further propose some all-optical devices based on CNTs, i.e., quantum optical transistor, optical Kerr modulator [Li (2011)] and mass sensor [Li (2012); Li and Zhu (2012)]. With regard to the CNT based mass sensor, the sensitivity can reach $2.3 \times 10^{-28} \mathrm{Hz} \cdot \mathrm{g}^{-1}$, which is five orders of magnitude more sensitive than the one using electric methods, and this can be used as the single atom mass sensing.

8.1 Theory

The model is based on a doubly clamped suspended carbon nanotube in the simultaneous presence of a strong pump field (with frequency ω_p) and a weak signal field (with frequency ω_s) as shown in Fig.8.2(b). Recently, such two-laser technique has been experimentally investigated by several groups to study the cavity optomechanical system and demonstrates the achievement of slow light and on-chip storage of light pulses [Safavi-Naeini (2011); Fiore (2011); Weis (2010); Teufel (2011)].

The localized exciton can be modeled as a two-level system consisting of the ground state $|g\rangle$ and the first excited state (single exciton) $|ex\rangle$ and can be characterized by the pseudospin $-1/2$ operators $S^{\pm}$ and S^z. The energy levels are shown in the inset window of Fig.8.2(b). Then the Hamiltonian of this localized two-level exciton can be described as $H_{\mathrm{ex}} = \hbar\omega_{\mathrm{ex}}S^z$, where ω_{ex} is the frequency of localized exciton. Besides, assuming the suspended carbon nanotube are characterized by sufficiently high quality factors, the lowest-energy resonance corresponds to the fundamental flexural mode with the frequency ω_n. The eigenmode of CNT can be described by a quantum harmonic oscillator, with b and b^+ the bosonic annihilation and creation operators of a quantum of energy $\hbar\omega_n$. Then the Hamiltonian for the vibration of nanotube resonator is given by $H_n = \hbar\omega_n b^+ b$.

In the simultaneous presence of a strong pump field and a weak signal field, the Hamiltonian of the carbon nanotube GOS can be written as [Wilson-Rae (2009); Boyd (2008)]

$$H = H_{ex} + H_n + H_{ex-n} + H_{ex-p}, \tag{8.1}$$

$$H_{ex-n} = \hbar\omega_n \beta S^z(b^+ + b), \tag{8.2}$$

$$H_{ex-p} = -\mu(S^+ E_p e^{-i\omega_p t} + S^- E_p^* e^{i\omega_p t})$$
$$- \mu(S^+ E_s e^{-i\omega_s t} + S^- E_s^* e^{i\omega_s t}), \tag{8.3}$$

where H_{ex-p} describes the localized exciton coupling to the two optical fields, E_p and E_s are slowly varying envelope of the pump field and signal field, respectively, μ is the electric dipole moment of the localized exciton. H_{ex-n} corresponds to the interaction between localized exciton and CNT oscillation mode [Wilson-Rae (2009); Graff (1991)], β is the coupling strength of the localized exciton-resonator, which can be given by [Wilson-Rae (2009)]

$$\beta = -2^{3/4}(1+\sigma)\frac{g\sigma_G^{1/4}\xi\varepsilon_\perp}{R(Eh)^{3/4}(q_0 L)}\sqrt{\frac{L}{\pi\hbar}}\cos 3\theta, \tag{8.4}$$

where σ, g, σ_G, E, R, q_0, $\sqrt{L}$, $\varepsilon_\perp$ and ξ represent the Poisson ratio, the off-diagonal deformation potential, the mass density of the CNT, the CNT Young modulus, the CNT radius, the thin rod elasticity phonon wavevector for the resonator mode, a direct consequence of the quadratic flexural dispersion, the electric field induced by gate electrodes along the carbon nanotube axis, and the intrinsic relative permittivity normal to the CNT axis, respectively. Here, we focus on the case of $\beta < 0.2$.

In a frame rotating at the pump field frequency ω_p, the total Hamiltonian of the suspended carbon nanotube can be expressed as:

$$H = \hbar\Delta_p S^z + \hbar\omega_n b^+ b + \hbar\omega_n \beta S^z(b^+ + b) - \hbar(\Omega S^+ + \Omega^* S^-)$$
$$-\mu(S^+ E_s e^{-i\delta t} + S^- E_s^* e^{i\delta t}), \tag{8.5}$$

where $\Delta_p = \omega_{ex} - \omega_p$ is the pump-exciton detuning, $\delta = \omega_s - \omega_p$ is the pump-signal detuning, and $\Omega = \mu E_p/\hbar$ is the Rabi frequency of the pump field.

Applying the Heisenberg equations of motion for operators S^z, S^- and Q and introducing the corresponding damping and noise terms, we derive the quantum Heisenberg-Langevin equations as follows [Gardiner (2000); Walls (1994)]:

$$\frac{d}{dt}S^z = -\Gamma_1(S^z + \frac{1}{2}) + i\Omega S^+ - i\Omega^* S^- + \frac{i\mu E_s e^{-i\delta t}}{\hbar}S^+ - \frac{i\mu E_s^* e^{i\delta t}}{\hbar}S^-,$$

(8.6)

$$\frac{d}{dt}S^- = -(i\Delta_p + \Gamma_2)S^- - i\omega_n\beta QS^- - 2i\Omega S^z - \frac{2i\mu E_s e^{-i\delta t}}{\hbar}S^z + \hat{F}_e,$$

(8.7)

$$\frac{d^2}{dt^2}Q + \frac{1}{\tau_n}\frac{d}{dt}Q + \omega_n^2 Q = -2\omega_n^2\beta S^z + \hat{\xi},$$

(8.8)

where $Q = b^+ + b$, Γ_1 and Γ_2 are the spontaneous emission rate and dephasing rate of localized exciton, respectively. τ_n is the vibration lifetime of carbon nanotube [Ekinci (2005); Wilson-Rae (2009)], which is inversely proportional to the decay rate of vibration of carbon nanotube. $\hat{F}_e$ is the δ-correlated Langevin noise operator, which has zero mean $\langle\hat{F}_e\rangle = 0$ and obeys the correlation function $\langle\hat{F}_e(t)\hat{F}_e^+(t')\rangle \sim \delta(t - t')$. In order to specify the candidate phonon mode in an actual CNT system, we consider the resonant flexural branch, which has a free spectral range larger than the optical linewidth of the zero phonon line of the excitonic transition. We focus on the laser excitation which is near resonant with the lowest-frequency flexural vibration mode of CNT [Wilson-Rae (2009, 2008)].

The motion of carbon nanotube resonator is affected by thermal bath of Brownian and non-Morkovian process [Gardiner (2000); Giovannetti (2001)]. The quantum effects on the resonator are only observed in the limit of high quality factor, that obeys $Q = \omega_n/\gamma_n \gg 1$. The Brownian noise operator can be modeled as Markovian with the decay rate γ_n ($\gamma_n = 1/\tau_n$) of the resonator mode. Therefore, the Brownian stochastic force has zero mean value $\langle\hat{\xi}\rangle = 0$ that can be characterized as [Giovannetti (2001)]

$$\langle\hat{\xi}^+(t)\hat{\xi}(t')\rangle = \frac{\gamma_n}{\omega_n}\int\frac{d\omega}{2\pi}\omega e^{-i\omega(t-t')}\left[1 + \coth\left(\frac{\hbar\omega}{2k_B T}\right)\right].$$

(8.9)

Following standard methods from quantum optics, we derive the steady-state solution to Eqs.(8.6)-(8.8) by setting all the time derivatives to zero. They are given by

$$S_0^- = \frac{-2\Omega S_0^z}{(\Delta_p + \omega_n\beta Q_0) - i\Gamma_2}, Q_0 = -2\beta S_0^z,$$

(8.10)

where S_0^z is determined by Eq.(8.21) (see below). In Chapter 2, we have demonstrated that no matter the strong coupling or the weak coupling, the quantum Heisenberg-Langevin always works well. Here, go beyond weak coupling, we can rewrite each Heisenberg operator as the sum of its steady-state mean value and a small fluctuation with zero mean value as follows $S^- = S_0^- + \delta S^-$, $S^z = S_0^z + \delta S^z$, $Q = Q_0 + \delta Q$. When inserting these equations into the Langevin Eqs.(8.6)-(8.8), the nonlinear term $\delta Q \delta S^-$ can be neglected safely. Assuming the classical coherent optical fields are weak, the operators can be identified by their expectation values, while dropping the quantum and thermal noise terms [Weis (2010)]. Then the linearized Langevin equations can be written as:

$$\langle \delta \dot{S}^z \rangle = -\Gamma_1 \langle \delta S^z \rangle + i\Omega \langle \delta (S^-)^* \rangle - i\Omega^* \langle \delta S^- \rangle \rangle + \frac{i\mu E_s e^{-i\delta t}}{\hbar} \langle \delta (S^-)^* \rangle$$
$$- \frac{i\mu E_s^* e^{i\delta t}}{\hbar} \langle \delta S^- \rangle, \tag{8.11}$$

$$\langle \delta \dot{S}^- \rangle = -(i\Delta_p + \Gamma_2)\langle \delta S^- \rangle - i\omega_n \beta (\langle \delta S^- \rangle Q_0 + S_0^- \langle \delta Q \rangle) - 2i\Omega \langle \delta S^z \rangle$$
$$- \frac{2i\mu E_s e^{-i\delta t}}{\hbar} \langle \delta S^z \rangle, \tag{8.12}$$

$$\langle \delta \ddot{Q} \rangle + \frac{1}{\tau_n} \langle \delta \dot{Q} \rangle + \omega_n^2 \langle \delta Q \rangle = -2\omega_n^2 \beta \langle \delta S^z \rangle. \tag{8.13}$$

In order to solve Eqs.(8.11)-(8.13), we make the ansatz [Weis (2010); Boyd (2008)] $\langle \delta S^- \rangle = S_+ e^{-i\delta t} + S_- e^{i\delta t}$, $\langle \delta S^z \rangle = S_+^z e^{-i\delta t} + S_-^z e^{i\delta t}$ and $\langle \delta Q \rangle = Q_+ e^{-i\delta t} + Q_- e^{i\delta t}$. Upon substituting these equations to Eqs. (8.11)-(8.13) and working to the lowest order in E_s, but to all orders in E_p, we obtain the linear optical susceptibility S_+ in the steady state as the following solution

$$\chi^{(1)}(\omega_s)_{\text{eff}} = \mu \frac{S_+}{\epsilon_0 E_s} = \frac{\mu^2}{\epsilon_0 \hbar \Gamma_2} \chi^{(1)}(\omega_s) = \Sigma_1 \chi^{(1)}(\omega_s), \tag{8.14}$$

where $\Sigma_1 = \mu^2/\epsilon_0 \hbar \Gamma_2$, ϵ_0 is the dielectric constant of vacuum. The dimensionless linear optical susceptibility is given by

$$\chi^{(1)}(\omega_s) = \frac{w_0}{f_1(\delta_0)} \times \{2e_1 \Omega_R^2 (e_1 + \delta_0)(e_2 + \omega_{n0} \beta^2 w_0 \zeta)$$
$$- e_1 e_2 [2\Omega_R^2 (e_1 + \omega_{n0} \beta^2 w_0 \zeta) - e_1(2i + \delta_0)(e_1 + \delta_0)]\}. \tag{8.15}$$

In the same way, the nonlinear optical susceptibility can be obtained by

$$\chi^{(3)}(\omega_s)_{\text{eff}} = \frac{\mu S_-}{3\varepsilon_0 E_p^2 E_s^*} = \frac{\mu^4}{3\varepsilon_0 \hbar^3 \Gamma_2^3} \chi^{(3)}(\omega_s) = \Sigma_3 \chi^{(3)}(\omega_s), \tag{8.16}$$

where $\Sigma_3 = \mu^4/3\varepsilon_0 \hbar^3 \Gamma_2^3$, and the dimensionless nonlinear susceptibility is given by

$$\chi^{(3)}(\omega_s) = \frac{w_0}{f_2(\delta_0)} 2e_1(\delta_0 - 2i)(e_2 + \omega_{n0}\beta^2 w_0 \zeta^*), \qquad (8.17)$$

where $e_1 = \Delta_{p0} - \omega_{n0}\beta^2 w_0 + i$, $e_2 = \Delta_{p0} - \omega_{n0}\beta^2 w_0 - i$, and $w_0 = 2S_0^z$, $\Gamma_1 = 2\Gamma_2$, $\omega_{n_0} = \omega_n/\Gamma_2$, $\tau_{n0} = \tau_n\Gamma_2$, $\Omega_R = \Omega/\Gamma_2$, $\delta_0 = \delta/\Gamma_2$, $\Delta_{p0} = \Delta_p/\Gamma_2$. The function $f_1(\delta_0)$ and $f_2(\delta_0)$ are given by

$$f_1(\delta_0) = e_1 e_2 (e_2 - \delta_0)[2\Omega_R^2(e_1 + \omega_{n0}\beta^2 w_0 \zeta) - e_1(2i + \delta_0)(e_1 + \delta_0)]$$
$$- 2e_1^2\Omega_R^2(e_1 + \delta_0)(e_2 + \omega_{n0}\beta^2 w_0 \zeta), \qquad (8.18)$$

$$f_2(\delta_0) = e_2^2 (e_2 + \delta_0)[2\Omega_R^2(e_1 + \omega_{n0}\beta^2 w_0 \zeta^*) + e_1(\delta_0 - 2i)(e_2 - \delta_0)]$$
$$- 2\Omega_R^2 e_1(e_1 - 2i)(e_1 - \delta_0)(e_2 + \omega_{n0}\beta^2 w_0 \zeta^*). \qquad (8.19)$$

The auxiliary function is written as

$$\zeta(\delta_0) = \frac{\omega_{n_0}^2}{\omega_{n_0}^2 - i\delta_0/\tau_{n0} - \delta_0^2}, \qquad (8.20)$$

and the population inversion w_0 of the localized exciton is determined by the following equation

$$(w_0 + 1)[(\Delta_{p0} - \omega_{n0}\beta^2 w_0)^2 + 1] + 2\Omega_R^2 w_0 = 0. \qquad (8.21)$$

8.2 Coherent optical spectroscopy

The cubic Eq.(8.21) has three roots, which characterizes the feature of optical multistability [Lam (1991)]. Figure 8.3 plots the bistable behavior of suspended carbon nanotube, according to Eq.(8.21). Similar with Fig.6.2 in Chapter 6, the carbon nanotube shows the stable state and unstable state simultaneously with different pump-exciton detunings.

In order to show the transmission spectrum clearly, we choose the realistic carbon nanotube resonator with an ambient temperature 4.2K, for $(\omega_n/2\pi, \Gamma_2, \gamma_n, \beta)$ = $(725\text{MHz}, 310\text{MHz}, 0.8\text{MHz}, 0.17)$ [Wilson-Rae (2009)]. For CNT based optomechanical system, the excitonic resonance couples to the vibration modes of CNT via deformation potential that matches the cavity mode interacting with mechanical resonator via radiation pressure, which demonstrates "optomechanically induced transparency" or called phonon induced transparency (PIT). Here, the vibration modes of CNT can be treated as phonon modes. Figure 8.4(a) exhibits this "optomechanically induced transparency" in a suspended CNT based optomechanical system, where the pump field detuning is driven on its red

(a)   (b)

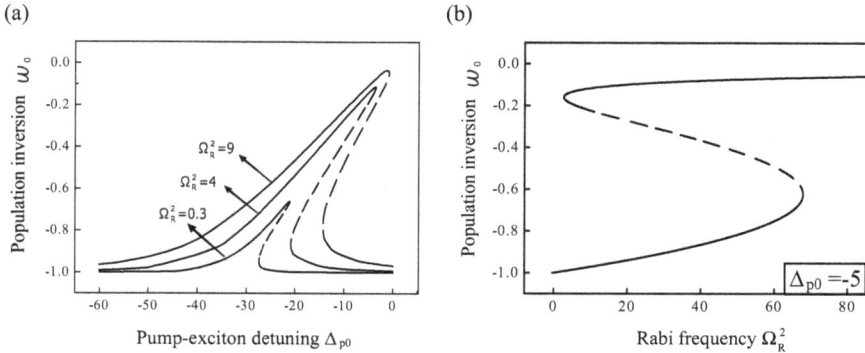

Fig. 8.3   The bistable behavior of carbon nanotube GOS. (a) The relationship between the population inversion w_0 and pump-exciton detuning, with different Rabi frequencies of pump field. (b) When fixing the pump-exciton detuning, the relationship between the population inversion w_0 and the Rabi frequency of pump field. Here, the dashed curves and solid curves correspond to the unstable state and stable state, respectively. The parameters used are $\beta = 0.8$ and $w_{n0} = 50$.

sideband, i.e., $\Delta_p = \omega_n$. Such an effect is very similar to the experimental results obtained by Kippenberg *et al.* [Weis (2010)]. However, if switching the pump-exciton detuning to $-\omega_n$ and increasing the pump field intensity gradually, the signal transmission exhibits a deeper dip as shown in Fig.8.4(b), where the negative transmission of the signal laser increases with the pump intensity. This is the so-called electromagnetically induced absorption (EIA) effect [Lezama (1999)]. As the pump beam intensity increases even further, the system switches from EIA to parametric amplification (PA) [Mollow (1967)] resulting in the signal beam amplification (Fig.8.4(c)), which is the same as that in the conventional optomechanical system [Safavi-Naeini (2011)]. Figure 8.4 gives us a specific idea that, when the intensity of pump laser increases from 0 to a larger value, the transmission of the signal laser presents PIT, EIA and PA effects, respectively, which is the second hallmark of cavity optomechanical system.

According to Eq.(8.14), we plot the resonance absorption spectrum of signal field with a carbon nanotube resonator, as shown in Fig.8.5(a). Since the original energy levels of the localized exciton have been dressed by the vibration modes of the nanotube in Fig.8.5(b), the uncoupled energy levels ($|ex\rangle$ and $|g\rangle$) split into dressed states $|ex, n\rangle$ and $|g, n\rangle$, respectively. We notice that the plot in Fig.8.5(a) displays three prominent features-the middle part consisting of three weak peaks and two steep peaks at the both sides, which are same with Fig.6.3 in Chapter 6. The middle

(a)

(b)

(c)

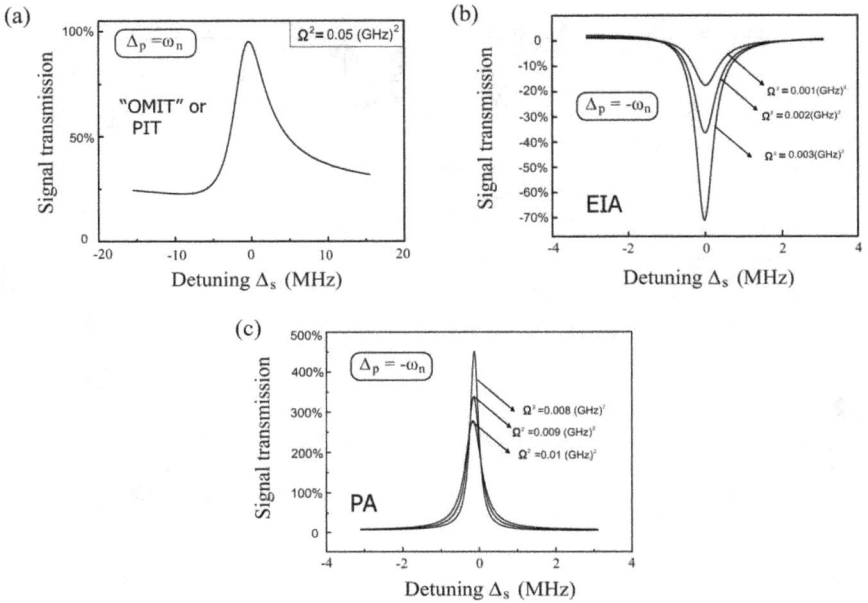

Fig. 8.4 Plots of optomechanically induced transparency (OMIT) or phonon induced transparency (PIT), electromagnetically induced absorption (EIA) and parametric amplification (PA). Signal transmission as a function of signal-exciton detuning for (a) $\Delta_p = \omega_n$. (b) $\Delta_p = -\omega_n$ with small Rabi frequencies of the pump field. (c) $\Delta_p = -\omega_n$ with large Rabi frequencies of the pump field.

part shows the origin of nanotube vibration induced stimulated Rayleigh resonance. Here the electrons make a transition from the lowest dressed level $|g, n\rangle$ to the dressed level $|ex, n\rangle$. Besides, the two steep peaks at the both sides represent the resonance amplification and absorption of the signal field after applying a strong pump field on the carbon nanotube resonator. The left peak corresponds to the amplification of the signal laser, where electrons make a transition from the lowest dressed level $|g, n\rangle$ to the highest dressed level $|ex, n + 1\rangle$ by the simultaneous absorption of two pump laser photons and emission of a photon at $\omega_p - \omega_n$. This process can amplify a wave at $\delta = -\omega_n$. Otherwise, the right peak is an absorption process, which corresponds to the usual absorption resonance as modified by the ac-Stark effect. Here coupling to phonons seems to provide the exciton with additional energy levels to realize EIT phenomena. Therefore in our structure one can obtain the slow output light without absorption by fixing $\Delta_p = 0$.

(a)

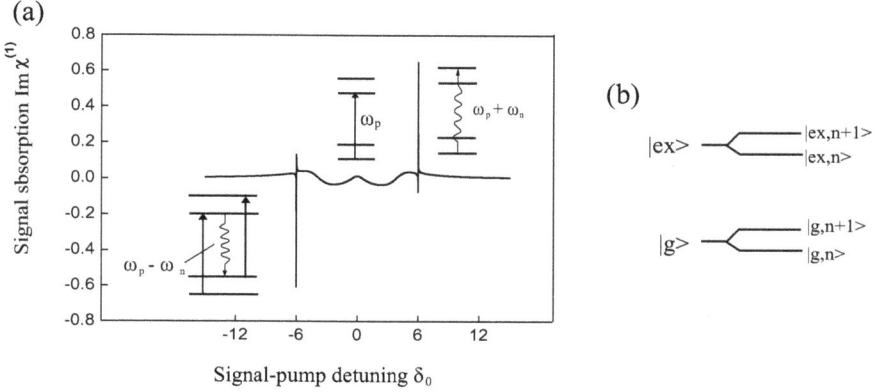

(b)

Fig. 8.5   (a) The absorption spectrum of a signal field as a function of signal-pump detuning for the case $\Omega_R^2 = 2$, $\omega_{n0} = 6$, $\Delta_{p0} = 0$, $\gamma_{n0} = 0.003$, and $\beta = 0.17$. Here the vibration modes is the same as the phonon modes. (b) The energy levels of localized exciton while dressing with the vibrational modes of nanotube resonator. $|n\rangle$ denotes the number state of the resonance mode.

8.3   Slow light and superluminal light

As usual, the signal light group velocity in a carbon nanotube resonator can be determined as [Bennink (2001); Harris (1992)]

$$v_g = \frac{c}{n + \omega_s(dn/d\omega_s)}, \tag{8.22}$$

where $n \approx 1 + 2\pi\chi_{\text{eff}}^{(1)}$, and then

$$\frac{c}{v_g} = 1 + 2\pi\text{Re}\chi_{\text{eff}}^{(1)}(\omega_s)_{\omega_s=\omega_{\text{ex}}} + 2\pi\omega_s\text{Re}(\frac{d\chi_{\text{eff}}^{(1)}}{d\omega_s})_{\omega_s=\omega_{\text{ex}}}. \tag{8.23}$$

We notice that when $\text{Re}\chi(\omega_s)_{\omega_s=\omega_{\text{ex}}}$ is zero and the dispersion is steeply positive or negative, the group velocity v_g is significantly reduced or increased. Here, the group velocity index n_g can be defined as

$$n_g = \frac{c}{v_g} - 1 = \frac{c - v_g}{v_g} = \frac{2\pi\omega_{\text{ex}}\rho\mu^2}{\hbar\Gamma_2}\text{Re}(\frac{d\chi^{(1)}(\omega_s)}{d\omega_s})_{\omega_s=\omega_{\text{ex}}}$$

$$= \Gamma_2\Sigma\text{Re}(\frac{d\chi^{(1)}(\omega_s)}{d\omega_s})_{\omega_s=\omega_{\text{ex}}}, \tag{8.24}$$

where $\Sigma = 2\pi\omega_{\text{ex}}\rho\mu^2/\epsilon_0\hbar\Gamma_2^2$. Then, the $n_g > 0$ and $n_g < 0$ correspond to the slow light and superluminal light, respectively [Boyd (2009)].

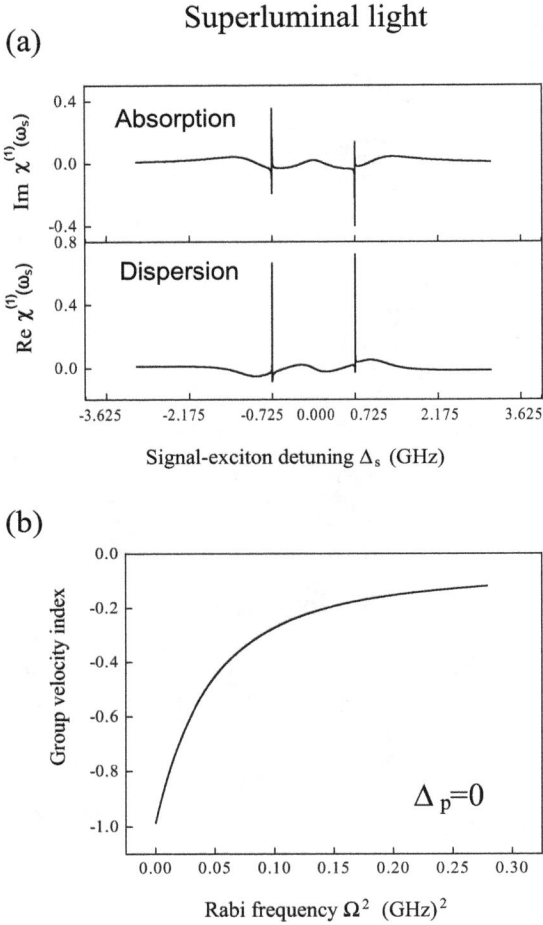

Fig. 8.6   Plots of superluminal light in a carbon nanotube resonator. (a) The dimensionless imaginary part and real part of the linear optical susceptibility (in units of Σ_1) as a function of the signal-exciton detuning. The parameters used are $\Delta_p = 0$ $\Omega^2 = 0.1(\text{GHz})^2$, $\omega_n = 0.725\text{GHz}$, $\gamma_n = 0.8\text{MHz}$, and $\beta = 0.17$. (b) The group velocity index n_g of superluminal light (in units of Σ) as a function of pump Rabi frequency Ω^2.

Figure 8.6(a) shows the imaginary part and real part of linear optical susceptibility while fixing $\Delta_p = 0$, which corresponds to the absorption and dispersion of the signal field, respectively. Here, we notice that the imaginary part has a zero absorption and the real part has a negative steep slope at $\Delta_s = 0$, which signifies the potential of superluminal light achievement. Since there is a negative slope of n and vanished absorption at $\Delta_s = 0$, it is possible to achieve the superluminal light in a carbon nanotube resonator. Figure 8.6(b) plots the group velocity index of signal laser n_g as a function of the Rabi frequency Ω^2, in the unit of Σ. Figure 8.6(b) shows that the output signal pulse can be about 10 times faster than input signal pulse in vacuum simply via tuning the pump laser on the resonant with exciton frequency in CNT resonator($\Delta_p = 0$).

Furthermore, there is an interesting feature we should notice, that the abscissa of the two steep peak shown in Fig.8.6(a) just corresponds to the vibrational frequency of CNT resonator ($\Delta_s = \omega_n = 725$MHz). That is, by fixing the pump laser on the exciton frequency $\Delta_p = 0$ and scanning the exciton frequency using another signal laser, one can obtain the vibrational frequency of CNT resonator either in signal absorption spectrum or in signal dispersion spectrum. This is a precise and easy method to get the vibrational frequency of CNT resonator in all-optical domain.

Furthermore, in the case of pump-off resonant ($\Delta_p = \omega_n$), the imaginary part and real part of linear optical susceptibility exhibit zero absorption and positive steep slope at $\Delta_s = 0$ in Fig.8.7(a), which denotes the possibility of ultraslow signal light. Figure 8.7(b) exhibits the slow light curve, where the most slow-light index can be produced in CNT resonator device as 180 as $\Omega^2 = 0.02$(GHz)^2. That is, the output signal pulse will be 180 times slower than the input light with a single CNT resonator.

Here, the magnitudes of superluminal light and ultra-slow light are determined by the number density of CNT resonator ρ. The physical origin of this result can also explained by the coupling between localized exciton and CNT vibration, which makes quantum interference between the CNT and the two optical fields via the localized exciton as $\delta = \omega_n$. Such a nonlinear process is due to phonon induced transparency (PIT), which has been discussed in Fig.8.4(a).

Figure 8.8(a) displays the the imaginary part of $\chi^{(1)}$ as a function of Δ_s for three different decay rates of γ_n. Figure 8.8(b) shows the amplification of the most remarkable region of transparency. Here, the linewidth of the signal spectrum increases as the decay rate γ_n increases. Therefore the shorter the CNT resonator decay is, the narrower of the signal spectrum linewidth is. Particularly, when the decay rate of the resonator is 0.1GHz, the linewidth of transparency window in Fig.8.8 becomes flat. As a result, the CNT resonator with small decay rate is beneficial to the transparency window. Here, we notice that the lifetime of carbon nanotube is $1/\gamma_n = 1.25\mu s$, which is three times larger than the dephasing

(a)

Slow light

(b)

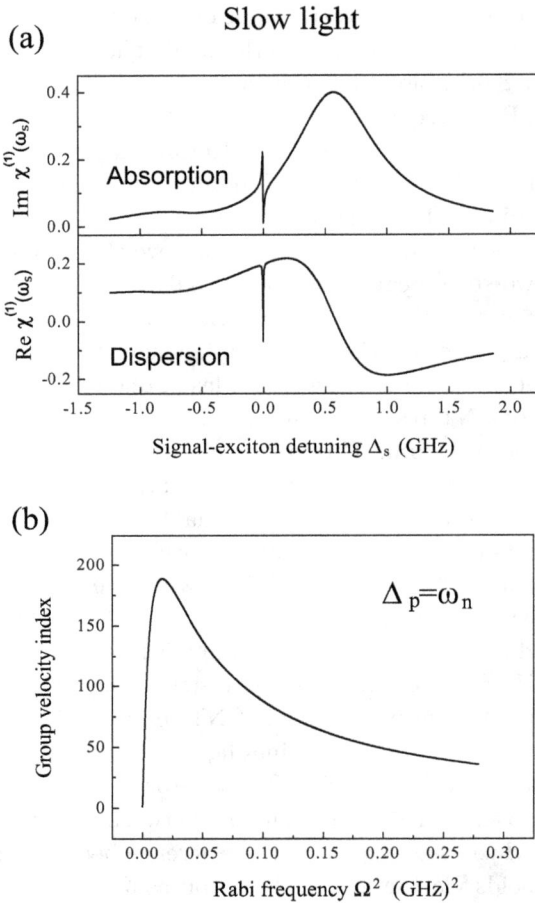

Fig. 8.7 The slow light using a carbon nanotube resonator. (a) The dimensionless imaginary part and real part of the linear optical susceptibility (in units of Σ_1) as a function of the signal-exciton detuning for $\Delta_p = \omega_n$; (b) The group velocity index n_g of slow light (in units of Σ) as a function of pump Rabi frequency Ω^2. The other parameters are the same with Fig.8.6.

time of localized exciton $1/\Gamma_2 = 0.003\mu s$. As discussed before, this is the necessary requirement for phonon induced transparency. Due to the high quality factor and short decay rate of CNT resonator, the superluminal light and ultra-slow light effect performed in CNT is obviously better than that in other quantum systems such as quantum wells and quantum dots [Zhu (2001)].

(a)

(b)

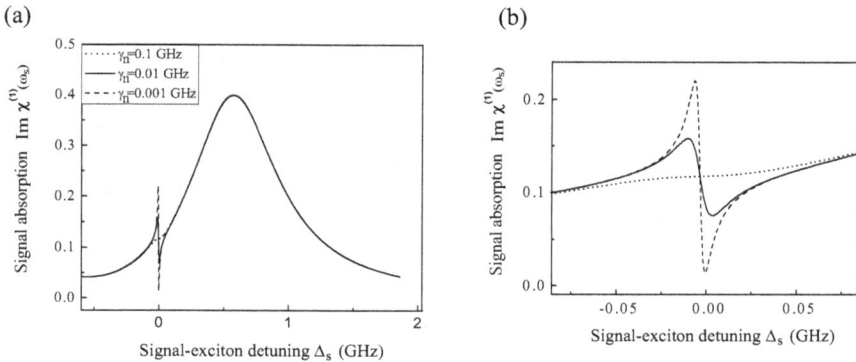

Fig. 8.8   (a) The absorption spectrum of a signal field as a function of the detuning Δ_s with three different decay rates of CNT resonator. The parameters used are $\Omega^2 = 0.1(\text{GHz})^2$, $\omega_n = 0.725\text{GHz}$, $\Delta_p = 0.725\text{GHz}$, and $\beta = 0.17$. (b) The amplification of the most remarkable region of transparency shown in (a).

8.4   Quantum optical transistor

Figure 8.9(a) shows the switching behavior of CNT resonator under the radiation of two optical fields. The top curve in Fig.8.9(a) displays the transmission spectrum when removing the pump beam but only applying a signal beam, which shows that in the absence of the pump beam a suspended CNT attenuates the weak signal beam totally. This dip arises from the usual exciton absorption resonance. However, as the pump beam turned

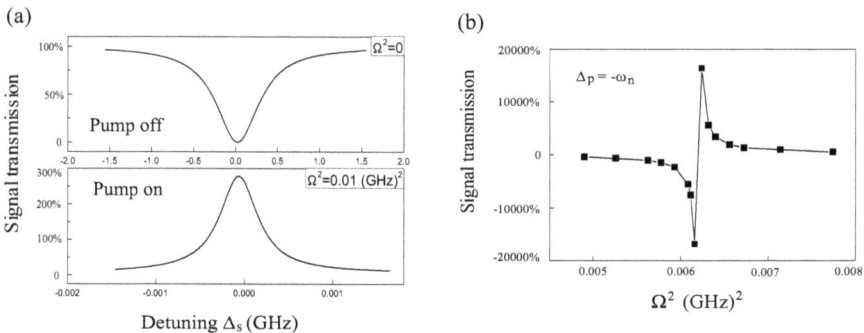

(a)

(b)

Fig. 8.9   (a) Attenuation (pump off) and amplification (pump on) of the signal beam, which indicates the switching behavior of carbon nanotube based optomechanical system. (b) The relationship between the signal transmission and the intensity of the pump laser. The parameters used are $(\omega_n/2\pi, \Gamma_2, \gamma_n, \Delta_p) = (725, 310, 0.8, -725)\text{MHz}$, $\beta = 0.17$.

on ($\Omega^2 = 0.01(\text{GHz})^2$), and fix the pump-exciton detuning on the blue side-band, i.e., $\Delta_p = -\omega_n$, the dip switches to a transmission peak immediately (the bottom curve in Fig.8.9(a)), which demonstrates that the output signal light is three times larger than the incident signal. This amplification corresponds to the negative absorption in Fig.8.5(a), which comes from the quantum interference between the CNT vibration modes (phonon modes) and the beat of the two optical fields via the localized exciton. Because of dressing with the phonon modes, the original two levels of exciton split into several metastable levels, then the electrons make transitions between them after applying a strong pump laser, which results in the constructive interference, and eventually amplifies the signal laser.

Figure 8.9(b) shows the relationship between the signal transmission and the Rabi frequencies of pump field. The curve in Fig.8.9(b) displays that there is a turning point between $\Omega^2 = 0.006(\text{GHz})^2$ and $\Omega^2 = 0.0065(\text{GHz})^2$, which changes the signal transmission from EIA to PA. It can be seen that the output signal laser can be 200 times larger than the incident pulse at $\Omega^2 = 0.0063(\text{GHz})^2$, which means the optomechanical system with a suspended CNT can not only switch the laser from off to on, but also serve as a laser amplifier via exciton-phonons interaction.

Figure 8.9 illustrates the switching behavior and amplification behavior in a doubly clamped carbon nanotube resonator, which provides an evidence that in the region $\Omega^2 \geq 0.0063 \ (\text{GHz})^2$ the optomechanical system based on CNT can act as a quantum optomechanical transistor under the radiation of two optical fields. Turning on and off the pump laser result in the amplification and attenuation of the signal laser, respectively, which declare that the pump laser acts like a signal controller while the CNT resonator acts like a carrier transport channel. Here, the switching time is decided by the dephasing time of localized exciton.

8.5 Nonlinear optical Kerr modulator

Current development of all-optical modulators and logic devices is highly expected for future optical information processing and telecommunication [Subramanian (2010); Kamaraju (2009)]. One of the best ways to implement ultrasensitive modulator is considered to make use of instantaneous optical Kerr effect (OKE) due to nonlinear refractive index [Zhou (2010); Zhong (2010)]. This effect reveals signatures of quantum dynamics, and has future applications in feedback control and adaptive measurements

[Jacobs (2007)]. In this section, we propose and analyze an optical Kerr modulator based on a carbon nanotube resonator, in the presence of a strong pump field and a weak signal field. In this scheme, the vibration of the nanotube makes a great contribution so that the optical Kerr effect can be enhanced significantly. Tuning the intensity of the pump field, the optical Kerr effect can be modulated instantaneously.

On the blue sideband, i.e., $\Delta_p = -\omega_n$, Fig.8.10(a) plots the enhanced optical Kerr coefficient $\mathrm{Re}\chi_{\mathrm{eff}}^{(3)}$ (dashed curve) and nonlinear absorption $\mathrm{Im}\chi_{\mathrm{eff}}^{(3)}$ (dotted curve) as a function of the detuning Δ_s, respectively. At $\Delta_s = 0$, the dotted curve in Fig.8.10(a) shows the negative nonlinear

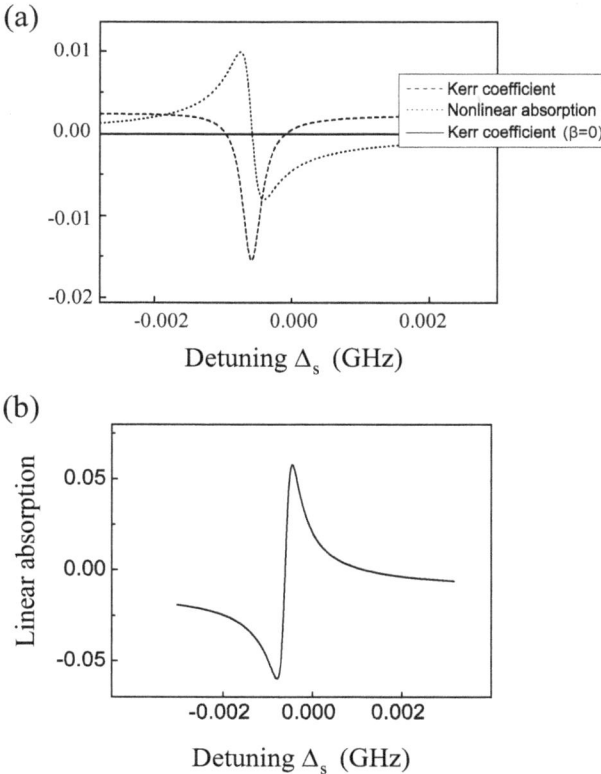

Fig. 8.10 (a) The optical Kerr coefficient and nonlinear absorption spectrum (in the unit of Σ_3) as a function of Δ_s. (b) The linear absorption coefficient (in the unit of Σ_1) as a function of detuning Δ_s. The parameters used are $\Omega^2 = 1.5(\mathrm{GHz})^2$, $\omega_n/2\pi = 725\mathrm{MHz}$, $\Delta_p = -725\mathrm{MHz}$, $\tau_n = 1.25\mu s$, $\beta = 0.17$ and $\Gamma_2 = 0.31\mathrm{GHz}$.

absorption. Besides, with the same parameters, Fig.8.10(b) displays vanished linear absorption ($\mathrm{Im}\chi_{\mathrm{eff}}^{(1)}$). It means that by fixing the pump detuning $\Delta_p = -\omega_n$ and scanning the second signal light across the exciton frequency, the large enhanced optical Kerr effect can be obtained at $\Delta_s = 0$. However, when ignoring the vibration of carbon nanotube, i.e., $\beta = 0$, the optical Kerr effect will disappear immediately (the solid curve in Fig.8.10(a)). Therefore, the coupling strength between the vibration of CNT and the localized exciton plays an important role for nonlinear Kerr effect in carbon nanotube optomechanical system.

Furthermore, increasing the pump power, the optical Kerr effect will be enhanced significantly (Fig.8.11(a)). In this case, the magnitude of nonlinear optical Kerr effect can be modulated by the pump laser. For example, the optical Kerr coefficient can be $\mathrm{Re}\chi_{\mathrm{eff}}^{(3)}(\omega_s) = 4 \times 10^{-14}\mathrm{m}^2/\mathrm{V}^2$, providing that $N = 10^{15}$ and $\mu = 10$ Debye [Datsyuk (2004)], when the

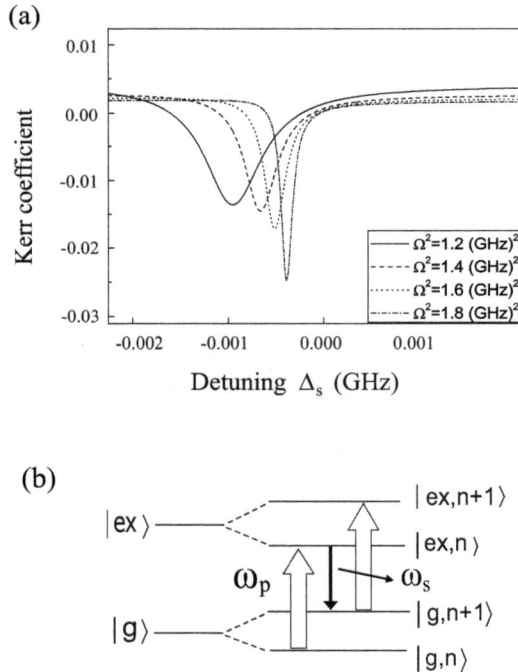

Fig. 8.11 The optical Kerr coefficient (in the unit of Σ_3) as a function of the detuning Δ_s for four different Rabi frequencies. (b) The energy levels and transitions shown in (a). The parameters used are the same with Fig.8.10.

Rabi frequency $\Omega^2 = 1.8$GHz. Figure 8.11(b) shows the energy levels and transitions of enhanced Kerr effect in (a), where the vibration wave of the nanotube resonator and the beat of the two optical fields become quantum interfered via the localized exciton that amplify the signal light at $\Delta_s = 0$.

Figure 8.12 shows the absolute value of optical Kerr coefficient as a function of Rabi frequency, which means that on the blue detuning of pump field ($\Delta_p = -\omega_n$), the magnitude of nonlinear optical Kerr coefficient of carbon nanotube can be modulated by the pump laser.

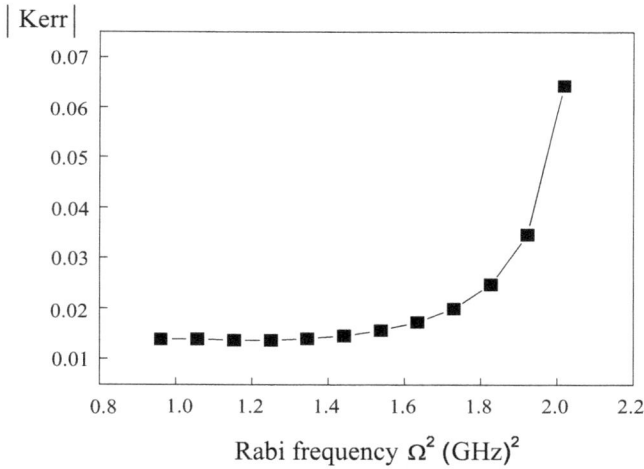

Fig. 8.12 The relationship between Kerr coefficient and the Rabi frequencies of pump field. Parameters used are the same with Fig.8.10.

Next, we demonstrate that the vibrational frequency of carbon nanotube can be measured in the nonlinear optical regime. Figure 8.13 plots the optical Kerr coefficient as a function of signal detuning Δ_s when $\Delta_p = 0$. In this figure, the two steep peaks locating at $\pm\Delta_s$ just correspond to the vibration frequency of carbon nanotube. In this case, we can propose a scheme to detect the frequency of nanotube resonator using this optical Kerr modulator. Here, if the frequency of nanotube resonator is $\omega_n/2\pi = 0.725$GHz, then the two peaks at ± 0.725GHz just match the resonator frequency (the solid curve in Fig.8.13). The dashed curve and dotted curve in Fig.8.13 correspond to different frequencies of nanotube resonator, where the location of the steep peaks also agree well with the resonator's frequency. This means that by fixing the pump field detuning $\Delta_p = 0$ and scanning

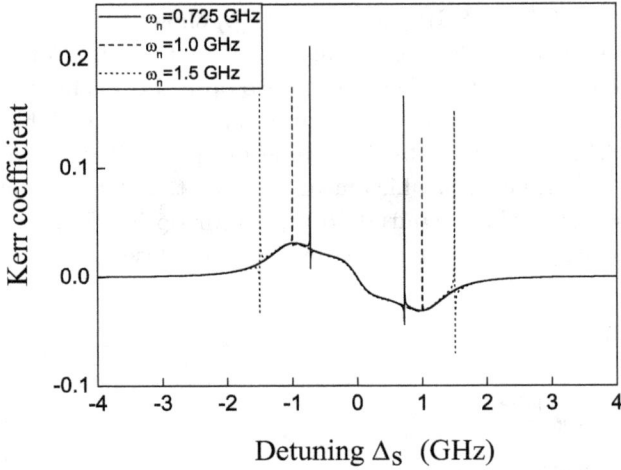

Fig. 8.13 The measurement of the frequency of nanotube resonator as a function of the detuning Δ_s for $\Omega^2 = 0.3(\text{GHz})^2$, $\Delta_p = 0$, $\tau_n = 1.25\mu\text{s}$, $\beta = 0.17$ and $\Gamma_2 = 0.31\text{GHz}$.

the signal field across the exciton frequency, the vibrational frequency of nanotube resonator can be detected in the signal absorption spectrum.

8.6   All-optical mass sensor with a carbon nanotube

Nanomechanical systems (NMS) are enabling important applications in diverse fields ranging from quantum measurement to biotechnology benefit from their sensitivity to external influences [Rugar (2004); Domon (2006); Gil-Santos (2009); Sinha (2006)]. Such enhanced sensitivity makes nanomechanical resonators including cantilever, doubly clamped beam and carbon nanotube emerge as strong candidates for mass sensors [Ekinci (2004); Yang (2006); Jensen (2008); Naik (2009)]. Carbon nanotube (CNT) is the best choice for ultrasensitive mass sensor among these numerous potential candidates. CNT based nano-sensors have the advantages that their sizes are thousands of times smaller than the nanoelectromechanical sensors' and the masses of carbon nanotubes ($\sim 10^{-21}\text{kg}$) are typically at least four orders of magnitude less than that of micromachined resonators($\sim 10^{-16}\text{kg}$). Even a tiny amount of atoms deposited onto the nanotubes will therefore make up a significant fraction of the total mass. On the other hand, the carbon nanotubes have two key material properties, low power consumption and high hardness, to push up the resonance frequency. They are also less sensitive to the variations of environmental temperature (compared to silicon piezoresistors) and have high specific area. Consequently, the CNT based nano-sensors are highly suitable for ultrasensitive mass sensor and biomedical sensor.

Carbon nanotube resonators act as mass sensors due to their resonant frequencies sensitive to the mass adsorbed to them. Even though the measurement techniques are rather challenging, the mass sensing principle remains simple. The carbon nanotube resonators can be described by harmonic oscillators with an effective mass m_n, a spring constant k, and a fundamental resonance frequency [Wilson-Rae (2009); Lassagne (2008)]

$$\omega_n = \sqrt{\frac{k}{m_n}}. \tag{8.25}$$

Mass sensing consists of monitoring the shift $\Delta\omega_n$ of ω_n induced by the adsorption onto the resonator of the molecular species to weigh. $\Delta\omega_n$ is related to the deposited mass m_d by [Ekinci (2004)]

$$m_d = (2\frac{m_n}{\omega_n})\Delta\omega_n, \tag{8.26}$$

where $\Re = (2m_n/\omega_n)^{-1}$ is defined as the mass responsivity. This is the direct relationship between external accreted mass and frequency-shift of carbon nanotube resonator. Therefore, if one can measure the frequency-shift in the signal absorption spectrum exactly, then the deposited mass can be obtained accurately.

In conventional electrical mass detection, the mechanical resonators should be suspended between two electrodes above a conducting plate, while a voltage applies through them [Sazonova (2004); Li (2004); Greaney (2009)]. The vibrational frequency is determined as $\omega_0/2\pi \approx \sqrt{E/\rho}(t/l^2)$, where t, l, E, ρ are the thickness, the length, the Young's modulus and the mass density of the NR, respectively. However, most of these electrical approaches require the device to be operated at high magnetic fields and at low temperatures. As the need for more ultrasensitive and precise methods to detect individual molecule mass such as virus particle, DNA and protein, incomplete technologies still exist for electrical mass detection. Firstly, in most nanomechanical resonators, especially in bilayer or multilayer structures, the internal strains must be taken into account when measuring resonant frequency, which will cause errors of mass detection due to the $\Delta\omega_n$-m_d [Ekinci (2005); Skyba (2010)]. Secondly, the electronic detection can not operate at the high frequency regions of nanomechanical resonator, which will decrease the mass sensitivity directly [Ekinci (2005); Schwab (2005)].

Addressed to this problem, by using all-optical technique, here we propose a scheme for mass sensing via a doubly clamped carbon nanotube resonator in all-optical domain. Owing to the advantages of carbon nanotube, this mass sensor has the potential to break through the limitation of frequency restriction and enhance the sensitivity of mass sensing. As mentioned above, comparing with the nanomechanical resonator, the carbon nanotube is highly suitable for sensitive measurement because of its small mass and high vibrational frequency. By measuring the resonance frequency shift of the nanotube in the signal absorption spectrum, we can easily determine the mass of external particles landing onto the surface of nanotube in all-optical domain.

In order to show the nanotube mass sensor clearly, we choose the experimentally realistic parameters of the carbon nanotube resonator for $(\Gamma_1, \Gamma_2, \beta) = (0.1\text{GHz}, 0.05\text{GHz}, 0.17)$ [Wilson-Rae (2009)]. The effective mass of carbon nanotube $m_n = 1580\text{zg}$, the vibrational frequency $\omega_n = 725\text{MHz}$, the high quality factor $Q = 1000$, and the nanotube lifetime $\tau_n = Q/\omega_n = 1.4\mu\text{s}$ [Jensen (2008)]. Since measuring the vibrational frequency of carbon nanotube in Fig.8.5 and Fig.8.6(a), using an all-optical

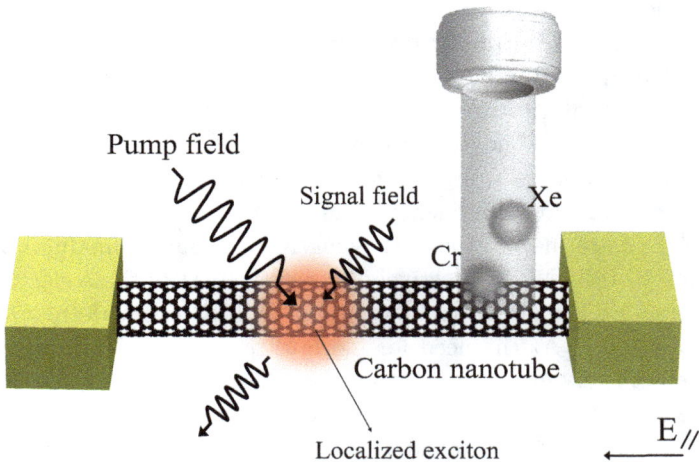

Fig. 8.14   Schematic diagram of mass sensor with a doubly clamped suspended carbon nanotube in the presence of a strong pump beam and a weak signal beam. The Chromium atoms or Xenon atoms are deposited onto the surface of the nanotube resonator in a special evaporator.

method. In the following, we shall illustrate how to weigh the mass of accreted particles landing onto the surface of carbon nanotube, according to Eq.(8.26). The physical layout of the entire nanotube mass sensor apparatus is shown in Fig.8.14. In general, all the experiments are done in situ within a cryogenically cooled, ultrahigh vacuum apparatus at ultra-low pressure.

The first step is to determine the original frequency of nanotube resonator. According to Eq.(8.26), we depict the signal absorption spectrum of carbon nanotube as a function of signal-exciton detuning Δ_s with $\Delta_p = 0$ (the dashed curve in Fig.8.15(a)). From this curve we find that there are two steep peaks at the both sides which just correspond to the frequency of carbon nanotube resonator. Particularly, for our selected nanotube the vibrational frequency is $\omega_n = 725$MHz, then the two steep peaks appear at ±725MHz have a linewidth of 120kHz (the right top window of Fig.15(a)). Such a phenomenon has been discussed in Fig.5. Actually, the amplitude of the middle feature is determined by the dephasing time of CNT exciton while the amplitudes of the two steep peaks at the both sides are decided by the vibrational lifetime of CNT. Here, the dephasing time of CNT exciton must be at least one order of magnitude smaller than the vibrational lifetime of CNT, due to mechanically induced coherent population oscillation. For the selected carbon nanotube, the vibrational lifetime ($1.4\mu s$) is much longer than the dephasing time of CNT exciton ($1/\Gamma_2 = 0.02\mu s$).

Next, we deposit samples onto the surface of carbon nanotube and measure the new vibrational frequency of CNT. Due to the relationship of Eq.(8.26), once measuring the frequency-shift in the signal spectrum, we can determine the mass of deposited particles. Here, we choose Cr atom as deposited samples, due to its active chemical nature and higher adsorption energy. For 1000 Cr atoms, the total mass is $m_d = 100$zg, which corresponds to $\Delta\omega_n = 10$MHz. The Cr solid curve in Fig.8.15(a) shows a slightly frequency shift from the original frequency of CNT.

Furthermore, it is well known that for the traditional mass spectrometry, the mass of noble metal atoms are very difficult to be determined due to the difficulty of its ionization. Therefore, the all-optical mass detection is necessary to be proposed, where the the noble metal atoms do not need to be ionized and their masses can be directly measured from the signal absorption spectrum conveniently. For example, if selecting 1000 Xe atoms ($m_d = 220$zg) landing onto the surface of CNT, the Xe curve in Fig.15(a) exhibits a large frequency-shift of $\Delta\omega_n = 23$MHz.

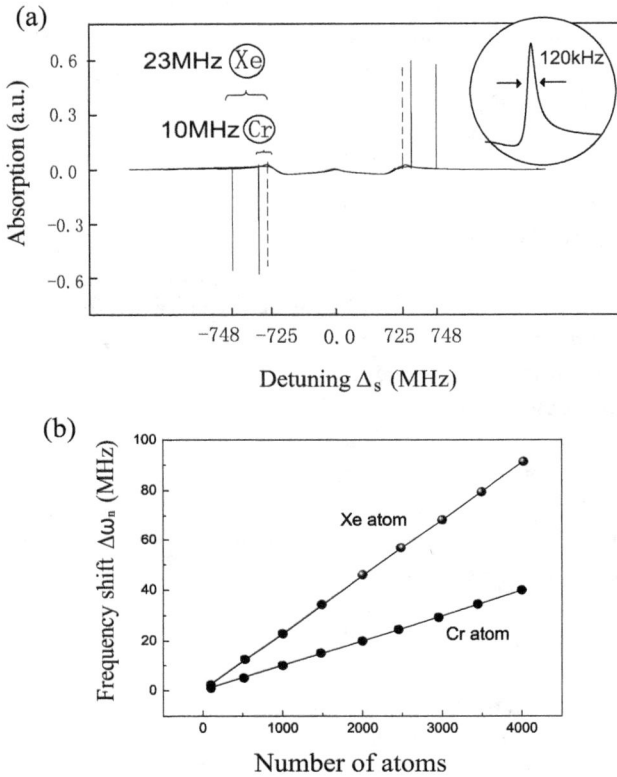

Fig. 8.15 (a) The signal absorption spectrum with and without landing the external atoms onto the surface of carbon nanotube resonator. The parameters used are $\Delta_p = 0$, $\tau_n = 1.4\mu s$, $\beta = 0.17$, $m_n = 1580zg$, $\omega_n = 725MHz$ and $\Omega^2 = 0.03(GHz)^2$. The dashed curve shows the initial resonance of carbon nanotube, while the other solid curves correspond to the frequency of CNT after landing Cr atoms ($\Delta m = 100zg$) and Xe atoms ($\Delta m = 220zg$), respectively. The right top corner curve shows the linewidth of absorption peak. (b) The relationship between the frequency-shift of CNT and numbers of deposited atoms.

Because of the small mass and high frequency of nanotube, the carbon nanotube based mass sensor has higher sensitivity. Figure 8.15(b) demonstrates the direct linear relationship between the resonance frequency shifts and the number of deposited atoms. The slope gives the mass sensitivity of the resonator. Clearly, smaller mass and higher frequency of the resonator enable higher mass sensitivity. During the theoretical analysis, the process of noise have not included, such as thermomechanical noise generated by the internal loss mechanisms in the resonator, and adsorption-desorption noise from residual gas molecules in the resonator packaging [Chiu (2008)], which will limit the performance for mass sensing of CNT. The CNT resonator environment includes a nonzero pressure of surface-contaminating molecules, which will be absorbed on a site of the resonator surface. Once these masses landing onto the surface of resonator, the resonator will have a very small frequency change which will cause a fractional frequency noise. This type of noise is the adsorption/desorption noise, which is not intrinsically dissipative. As the arrival and departure time of the molecules is random, they do not on average change the energy of the resonator and will leave the quality factor unchanged. Therefore, compared with the adsorption-desorption noise, the thermal noise of the mechanical motion is the dominant noise source. In this case, the induced frequency shift of CNT during the mass sensing experiment can be resolvable.

8.7   Surface plasmon enhanced optical mass sensor

In classical picture, plasmons can be described as the oscillation of free electron density against the fixed positive ions in a metal [Wang (2011)]. Just since a decade ago, scientific results have shown that by coating the metallic nanoparticles in the vicinity of quantum dot or quantum well materials, the surface plasmon produced by metallic nanoparticles after applying the optical fields, will enhance the emitting efficiency of QD/QW [Qiu (2011); Wang (2011); Zhmakin (2011)]. Surface plasmon can enhance the signals and amplify the field strengths in hybrid plasmonic nanostructures, which contributes to new properties of nonlinear response and light propagation [Zhang (2006, 2011)]. As early as 2008, we have investigated the light propagation in a single quantum dot coupled to metallic nanoparticle system. We demonstrated that the signal absorption spectrum can be greatly increased, due to the plasmonic enhancement effect [Lu (2008, 2009)]. Recently, the same idea and treatment have been used by Fofang *et al.* [Fofang (2011)] in laboratory and Cabrera-Granado in theory, respectively [Cabrera-Granado (2011)]. The former has studied the ultrafast optical dynamics of excitons in strongly coupled excitons and plasmons system, while the later is the case of study of slow-light performance of molecular aggregates arranged in nanofilms by means of coherent population oscillations. Both of them have demonstrated that the participation of surface plasmon provides enhanced optical properties and slow light effect.

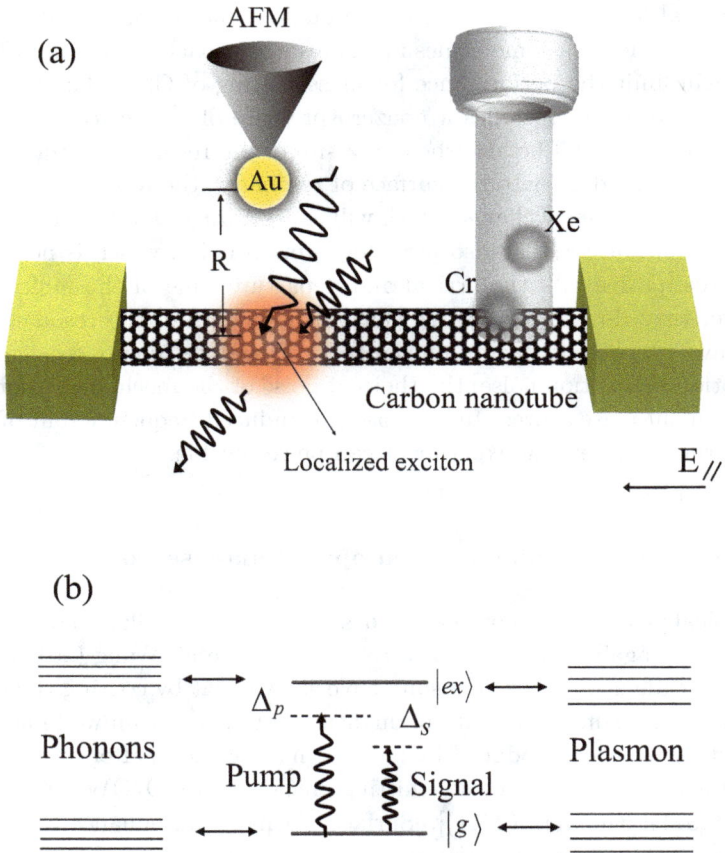

Fig. 8.16 (a) In the presence of a strong pump beam and a weak signal beam, a carbon nanotube suspended between two clamps is used as mass sensing. A single gold metal nanoparticle (MNP) attached to the end of a steep optical fibre tip is positioned above the CNT. The gold MNP has the radius a_0 and a center-to-center distance R towards the CNT. The Chromium atoms or Xenon atoms are deposited onto the surface of carbon nanotube. (b) The energy levels of localized exciton while dressing the surface plasmon and the vibration of carbon nanotube.

In this section, under the radiation of two optical fields, we propose an all-optical controlled mass sensing down to a single atom regime, via a hybrid system consisting of surface plasmon and doubly clamped carbon nanotube resonator. Compared with last section, here, a metallic nanoparticle placing in the vicinity of localized exciton of carbon nanotube is to enhance the signal absorption spectrum, and eventually enhance the sensitivity of mass sensing. The mass of a single atom is determined while it is landed onto the surface of nanotube in terms of the vibrational frequency-shift of carbon nanotube. Rather than the traditional electrical mass detection, the single atom mass sensing based on double optical-excitations has a narrower linewidth (kHz) and a higher sensitivity ($2.3 \times 10^{-28} \mathrm{Hz} \cdot \mathrm{g}^{-1}$), which is five orders of magnitude more sensitive than the one using electric methods, due to the advantages of carbon nanotube (i.e., ultra-light mass and high quality factor) and surface plasmon (spectrum enhancement).

The proposed mass sensor is based on a plasmon and a doubly clamped carbon nanotube in the simultaneous presence of a strong pump field (with frequency ω_p) and a weak signal field (with frequency ω_s) as shown in Fig.8.16. A single gold metal nanoparticle (MNP) attached to the end of a steep optical fibre tip is positioned above the CNT. An atomic force microscope (AFM) is used to probe the tip and to stabilize its distance [Sandoghdar (2006); Kalkbrenner (2004, 2005)]. The gold MNP has the radius a_0 and a center-to-center distance R towards the CNT. For a doubly clamped carbon nanotube resonator, the center of mass of the exciton is localized via the spatial modulation of the Stark-shift induced by a static inhomogenous electric field. Therefore, the carbon nanotube acts like a quantum dot when electrons are confined to a small region of carbon nanotube [Yasukochi (2011)], which has been discussed in the last section.

As usual, the localized exciton can be modeled as a two-level system consisting of the ground state $|g\rangle$ and the first excited state (single exciton) $|\mathrm{ex}\rangle$, which can be characterized by the pseudospin $-1/2$ operators $S^{\pm}$ and S^z. Then the Hamiltonian of this localized exciton can be described as $H_{\mathrm{ex}} = \hbar\omega_{\mathrm{ex}}S^z$, where ω_{ex} is the frequency of exciton. The Hamiltonian for the resonance of nanotube resonator is given by $H_n = \hbar\omega_n a^+ a$, where ω_n denotes the vibrational frequency of carbon nanotube, a and a^+ are the annihilation and creation operators. In the simultaneous presence of a strong pump field and a weak signal field, the Hamiltonian of the system can be written as [Wilson-Rae (2009); Zhang (2006); Yasukochi (2011); Graff (1991); Yan (2008)]

$$H = \hbar\omega_{\mathrm{ex}}S^z + \hbar\omega_n a^+ a + \hbar\omega_n \beta S^z(a^+ + a) - \mu(E_{LE}S^+ + E^*_{LE}S^-), \quad (8.27)$$

where β is the coupling strength of the CNT mode and localized exciton, μ is the dipole matrix element of the exciton. E_{LE} is the total optical field felt by the localized exciton, which can be expressed as

$$\tilde{E}_{LE} = E_p + E_s e^{-i\delta t} + \frac{S_\alpha P_{MNP}}{\varepsilon_{\mathrm{eff1}} R^3}, \quad (8.28)$$

$$P_{MNP} = \gamma a_0^3 (E_p + E_s e^{-i\delta t} + \frac{S_\alpha P_{LE}}{\varepsilon_{\text{eff2}} R^3}), \tag{8.29}$$

where P_{MNP} is the dipole which comes from the charge induced by the signal field [Zhang (2006); Yan (2008)], $\delta = \omega_p - \omega_s$ is the detuning of the signal and pump field. The dipole moment of the localized exciton is expressed via the off-diagonal elements of the density matrix: $P_{LE} = \mu S^-$ [Yariv (1975)]. The dipole approximation used here is reasonable when the distance R is large and the exciton-plasmon interaction is relatively weak [Yan (2008)]. For a spherical particle (such as Au atom) whose radius is much smaller than the wavelength of light, the electric field is uniform across the particle and the electrostatic (Rayleigh) approximation is a good one. S_α is polar factor for electric field polarization. Here, we set $S_\alpha = 2$, which corresponds to the polar direction is along the z axis of the hybrid system.

The other parameters are expressed as

$$\varepsilon_{\text{eff1}} = \frac{2\varepsilon_0 + \varepsilon_s}{3\varepsilon_0}, \tag{8.30}$$

$$\varepsilon_{\text{eff2}} = \frac{2\varepsilon_0 + \varepsilon_{\text{Au}}(\omega)}{3\varepsilon_0}, \tag{8.31}$$

$$\varepsilon_{\text{Au}}(\omega) = 1 - \frac{\omega_{\text{pl}}^2}{\omega(\omega + i\gamma_{\text{Au}})}, \tag{8.32}$$

where ε_0 and ε_s are the dielectric constants of the background medium and localized exciton, respectively; $\varepsilon_{\text{Au}}(\omega)$ is the Au nanoparticle's dielectric constant; ω_{pl} and γ_{Au} are the bulk metal plasma frequency and the frequency-dependent damping, respectively. The imaginary part of relative permittivity ε_{Au} determines the metallic losses [Pelton (2008); Khoo (2006)]. Therefore the total optical field E_{LE} felt by the localized exciton is

$$E_{LE} = A(E_p + E_s e^{-i\delta t}) + \mu B S^-, \tag{8.33}$$

where

$$A = 1 + \frac{\gamma a_0^3 S_\alpha}{\varepsilon_{\text{eff1}} R^3}, \tag{8.34}$$

$$B = \frac{\gamma a_0^3 S_\alpha^2}{\varepsilon_{\text{eff1}} \varepsilon_{\text{eff2}} R^6}, \tag{8.35}$$

$$\gamma = \frac{\varepsilon_{\mathrm{Au}}(\omega) - \varepsilon_0}{2\varepsilon_0 + \varepsilon_{\mathrm{Au}}(\omega)}. \tag{8.36}$$

In a rotating frame at the pump field frequency ω_p, the total Hamiltonian is given by

$$H = \hbar\Delta_p S^z + \hbar\omega_n a^\dagger a + \hbar\omega_n \beta S^z (a^\dagger + a) - \mu(\tilde{E}_{LE} S^\dagger + \tilde{E}_{LE}^* S^-), \tag{8.37}$$

where $\Delta_p = \omega_{\mathrm{ex}} - \omega_p$ is the frequency detuning between the exciton and pump field. Next, we use the density matrix approach described in chapter 2 to deal with the Hamiltonian (8.37), then

$$\frac{d\langle S^z \rangle}{dt} = -\Gamma_1(\langle S^z \rangle + \frac{1}{2}) + i\Omega(A^*\langle S^- \rangle - A\langle S^+ \rangle) + \frac{i\mu^2}{\hbar}$$

$$[\langle S^+ \rangle \langle S^- \rangle (B - B^*)] + \frac{i\mu}{\hbar}(A^* E_s^* \langle S^- \rangle e^{i\delta t} - A E_s \langle S^+ \rangle e^{-i\delta t}), \tag{8.38}$$

$$\frac{d\langle S^- \rangle}{dt} = -[\Gamma_2 + i(\Delta + \omega_n \beta \langle Q \rangle)]\langle S^- \rangle - 2i\Omega A \langle S^z \rangle$$

$$- \frac{2i\mu}{\hbar} A E_s e^{-i\delta t} - \frac{2i\mu^2 \langle S^z \rangle \langle S^- \rangle}{\hbar}, \tag{8.39}$$

$$\frac{d^2 \langle Q \rangle}{dt^2} + \gamma_n \frac{d\langle Q \rangle}{dt} + \omega_n^2 \langle Q \rangle = -2\beta\omega_n^2 \langle S^z \rangle, \tag{8.40}$$

where $\Omega = \mu E_p / \hbar$ is the Rabi frequency of the pump field, Γ_1 and Γ_2 denote the relaxation rate and dephasing rate of localized exciton, respectively. γ_n is the decay rate of the carbon nanotube resonator due to the coupling to a reservoir of "background" modes and the other intrinsic processes [Ekinci (2005); Wilson-Rae (2009)]. They are derived microscopically as

$$\Gamma_1 = \frac{2}{\hbar}\{\frac{\gamma_1}{2}(1 + 2N(\omega_{\mathrm{ex}}))\}, \tag{8.41}$$

$$\Gamma_2 = \frac{1}{\hbar}\{\frac{\gamma_1}{2}(1 + 2N(\omega_{\mathrm{ex}}))\} + \frac{4}{\hbar}\{\frac{\gamma_2}{2}(1 + 2N(0))\}, \tag{8.42}$$

$$\gamma_n = \frac{\gamma_3}{2m_n\omega_n}. \tag{8.43}$$

Obviously, when the pure dephasing coupling is neglected, i.e., $\gamma_2 = 0$, we can get $\Gamma_1 = 2\Gamma_2$. In order to solve these equations, we take the semiclassical approach by factorizing the nanotube and exciton degrees of freedom, i.e., $\langle QS^- \rangle = \langle Q \rangle \langle S^- \rangle$, in which any entanglement between these systems should be ignored. And then we make the ansatz [Boyd (2008)]:

$$\langle S^-(t) \rangle = S_0 + S_+ e^{-i\delta t} + S_- e^{i\delta t}, \tag{8.44}$$

$$\langle S^z(t) \rangle = S_0^z + S_+^z e^{-i\delta t} + S_-^z e^{i\delta t}, \tag{8.45}$$

$$\langle Q(t) \rangle = Q_0 + Q_+ e^{-i\delta t} + Q_- e^{i\delta t}. \tag{8.46}$$

Upon substituting the approximation to Eqs.(8.38)-(8.40)and working to the lowest order in E_s, but to all orders in E_p, we finally obtain the linear optical susceptibility S_+ in the steady state as the following solution

$$\chi_{\text{eff}}^{(1)}(\omega_s) = \frac{\mu}{\varepsilon_0 E_s} S_+ = \frac{\mu^2}{\varepsilon_0 \hbar \Gamma_2} \chi^{(1)}(\omega_s) = \Sigma \chi^{(1)}(\omega_s), \tag{8.47}$$

where $\Sigma = \mu^2/(\varepsilon_0 \hbar \Gamma_2)$, and the dimensionless linear susceptibility is given by

$$\chi^{(1)}(\omega_s) = \frac{2A^2 dw_0[(e + \Omega_0^2 A)(c + \delta_0) - B_0 \Omega_0^2 Aw_0] - Aw_0 g(c + \delta_0)}{cg(c + \delta_0) - 2d[A^*(d - \delta_0) + 2iB_{I0}Aw_0][(e + \Omega_0^2 A)(c + \delta_0) - B\Omega_0^2 Aw_0]}, \tag{8.48}$$

where

$$c(\delta_0) = \Delta_{p0} - \delta_0 - \omega_{n0}\beta^2 w_0 + B_{R0} w_0 - i(1 - B_{I0}w_0), \tag{8.49}$$

$$d(\delta_0) = \Delta_{p0} + \delta_0 - \omega_{n0}\beta^2 w_0 + B_{R0} w_0 + i(1 - B_{I0}w_0), \tag{8.50}$$

$$e(\delta_0) = \frac{\Omega_0^2 \omega_{n0}\beta^2 w_0 \zeta}{\Delta_{p0} - \omega_{n0}\beta^2 w_0 + B_{R0} w_0 - i(1 - B_{I0}w_0)}, \tag{8.51}$$

$$f(\delta_0) = \frac{\Omega_0^2 \omega_{n0}\beta^2 w_0 \zeta}{\Delta_{p0} - \omega_{n0}\beta^2 w_0 + B_{R0} w_0 + i(1 - B_{I0}w_0)}, \tag{8.52}$$

$$g(\delta_0) = 2[A(c + \delta_0) - 2iB_{I0}Aw_0][(f + \Omega_0^2 A)(d - \delta_0) - B_0^* \Omega_0^2 Aw_0] - id(\Gamma_{10} - i\delta_0)(c + \delta_0)(d - \delta_0). \tag{8.53}$$

The auxiliary function $\zeta(\omega_s)$ is given by

$$\zeta(\omega_s) = \frac{\omega_{n0}^2}{\omega_{n0}^2 - \delta_0^2 - i\gamma_{n0}\delta_0}, \tag{8.54}$$

where $w_0 = 2S_0^z$, $\delta_0 = \delta/\Gamma_2$, $\Omega_0 = \Omega/\Gamma_2$, $\omega_{n0} = \omega_n/\Gamma_2$, $\Delta_{p0} = \Delta_p/\Gamma_2$, $\Gamma_{10} = \Gamma_1/\Gamma_2$, $B_0 = \mu^2 B/(\hbar \Gamma_2)$, $B_{R0} = \text{Re}(B_0)$, and $B_{I0} = \text{Im}(B_0)$. The population inversion (w_0) of the localized exciton in CNT is determined by the equation

$$(w_0 + 1)[(1 - B_{I0}w_0)^2 + (\Delta_{p0} - \omega_{n0}\beta^2 w_0 + B_{R0}w_0)^2] + 2\Omega_0^2 A^2 w_0 = 0. \tag{8.55}$$

Here, in order to show single atom mass sensor clearly, we choose the realistic carbon nanotube resonator coupled to Au nanoparticle. The parameters used are $(\Gamma_1, \Gamma_2, \beta, a_0, R, \mu, \varepsilon_0, \varepsilon_s, \gamma_{\mathrm{Au}}) = (0.6\mathrm{GHz}, 0.3\mathrm{GHz}, 0.17, 2.5\mathrm{nm}, 18\mathrm{nm}, 40D, 1, 6, \omega_{\mathrm{pl}}/60)$ [Zhang (2006); Wilson-Rae (2009); Yan (2008); Pelton (2008)]. The effective mass of carbon nanotube $m_n = 1580\mathrm{zg}$, which corresponds to the vibrational frequency $\omega_n = 725\mathrm{MHz}$, the quality factor $Q = 1000$, and the decay rate $\gamma_n = \omega_n/Q = 8 \times 10^5 \mathrm{Hz}$ [Jensen (2008)].

Figure 8.16 shows our proposed setup of surface plasmon enhanced mass sensing in the presence of a strong pump laser and a weak signal laser. The two optical fields are aimed at the localized exciton of nanotube resonator.

We repeat the mass sensing process described in the last section, that is to measure the vibrational frequency of carbon nanotube after and before landing deposited samples, respectively. According to Eq.(8.26), the mass of deposited particle can be determined if measuring the frequency-shift of carbon nanotube precisely. According to Eq.(8.48), we first radiate a strong pump laser on the hybrid MNP-CNT system, and then detect the absorption spectrum of the second signal laser as a function of signal-exciton detuning Δ_s, with $\Delta_p = 0$. Figure 8.17(a) displays two steep peaks at the both sides, which just correspond to the vibrational frequency of carbon nanotube. Particularly, for the vibrational frequency $\omega_n = 725\mathrm{MHz}$, the two steep peaks will locate at $\Delta_s = \pm 725\mathrm{MHz}$, respectively. The two steep peaks represent the resonance amplification and absorption of the vibrational mode of nanotube, respectively. The energy levels and transitions are placed in the vicinity of each prominent peak. Here, since dressing with the vibration mode of CNT and plasmon mode, the original energy levels of localized exciton ($|ex\rangle$ and $|g\rangle$) split into dressed states $|ex, n\rangle$ and $|g, n\rangle$. The left amplified peak signifies the transition from the lowest dressed level $|g, n\rangle$ to the highest dressed level $|ex, n+1\rangle$, while the right absorbed peak shows the usual absorption resonance as modified by the ac-Stark effect. Besides, the middle feature is vibration induced stimulated Rayleigh resonance, which corresponds to a transition from the lowest dressed level $|g, n\rangle$ to the dressed level $|ex, n\rangle$.

Figure 8.17(b) shows the enlarged view of the right absorption peak shown in Fig.8.17(a), with and without considering the impact of plasmon mode. It is obvious that the role of plasmon during the mass sensing is to narrow and enhance the signal spectrum, which will increase the sensitivity of mass sensing. Different R corresponding to different linewidth offers an opportunity to reach the optimal sensitivity of mass sensing via control of

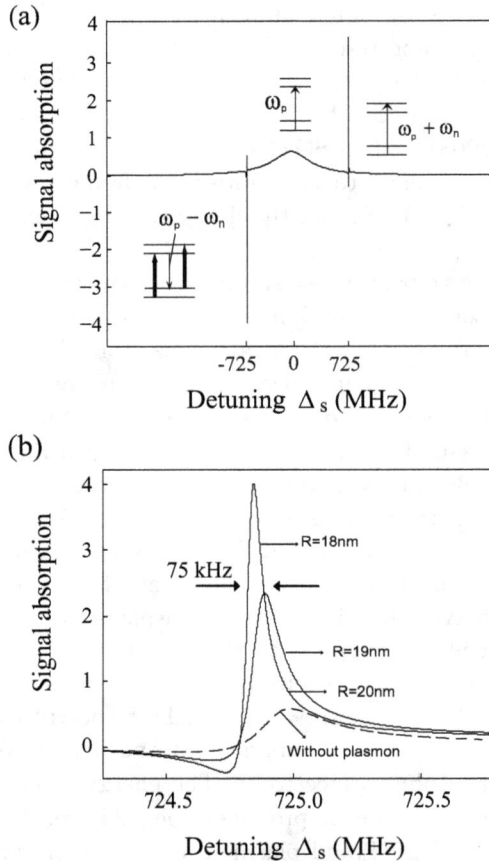

Fig. 8.17   (a) The signal absorption spectrum as a function of Δ_s. The inset corresponds to the transitions and energy levels, where $|n\rangle$ denotes the number states of the nanotube resonance mode. (b) The enlarged view of the right absorbtion peak shown in (a), with and without surface plasmon. The linewidth of signal laser is about 75kHz while fixing the distance between Au MNP and carbon nanotube on 18nm. The parameters used are $\Delta_p = 0$, $\gamma_n = 8 \times 10^5$ Hz, $\beta = 0.17$, $\omega_n = 725$ MHz and $\Omega^2 = 0.01(\text{GHz})^2$.

the distance between the Au MNP and the CNT resonator. When $R = 18$nm, the linewidth of signal absorption is ~ 75kHz, which demonstrates that the proposed mass sensor has enough spectral resolution to do the mass sensing in a single atom regime.

The next step is track the resonance frequency-shift of CNT before and after injecting sample. Here, we deposit a single atom onto the surface

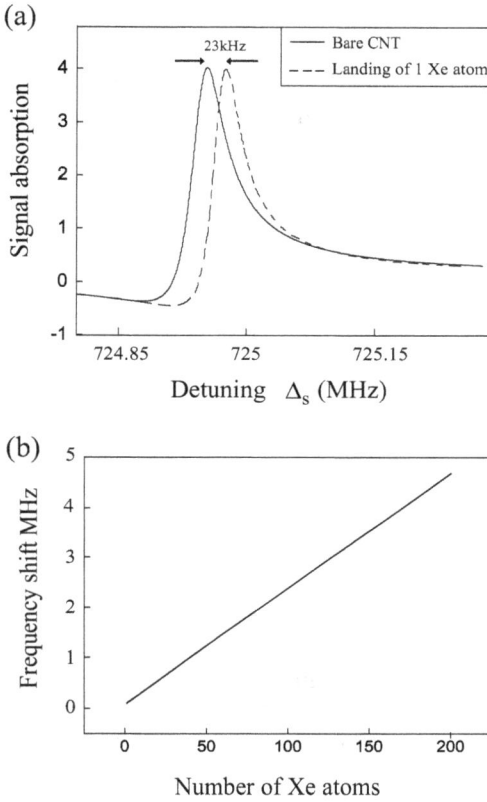

Fig. 8.18 (a) The signal absorption spectrum with and without landing a single Xe atom onto the surface of carbon nanotube. The solid curve and dashed curve display the vibrational frequency of CNT before and after landing a single Xe atom. (b) The resonance frequency shift as a function of the atom numbers. The parameters used are $\Delta_p = 0$, $\gamma_n = 8 \times 10^5 \text{Hz}$, $\beta = 0.17$, $m_n = 1580\text{zg}$, $\omega_n = 725\text{MHz}$ and $\Omega^2 = 0.01(\text{GHz})^2$.

of carbon nanotube, then the spectral peaks representing the resonant frequency of CNT will be shifted according to the equation $\Delta m = (2m_n\Delta\omega_n)/\omega_n$. For a single Xe atom with the mass $\Delta m = 0.22\text{zg}$, the dashed curve in Fig.8.18(a) exhibits a frequency-shift of $\Delta\omega_n = 23\text{kHz}$. Figure 8.18(b) demonstrates the direct linear relationship between the frequency shifts and the number of atoms landing on the resonator, which shows that the single atom weighing scheme is also suitable in multiple atoms detection. The slope gives the mass responsivity of the resonator [Lassagne (2008)] $\Re = |(2m_n/\omega_n)^{-1}| = 2.3 \times 10^{-28}\text{Hz} \cdot \text{g}^{-1}$, which is five orders of magnitude higher than that in conventional mass measurement under electric environment.

Since the deposited atoms only change the vibrational frequency of the CNT, the mass sensing proposed here can also apply to weigh chemically active atoms. Here, the size and the electronic structure of localized exciton are determined by the external voltage. It is well know that the size of chemically active atom is less than 1nm, which is very smaller than the length of nanotube (~ 120nm). Assuming the effective size of localized exciton is around 2nm [Chang (2004); Capaz (2006); Malic (2010)], and fixing the adsorbed site far away from the localized exciton, the size of perturbation on electronic state of CNT while depositing chemically active atom is hardly recognized. In this case, the hybridization of electronic states of CNT induced by chemisorbing active atoms can be neglected safely.

Bibliography

A. N. Khlobystov, Carbon nanotubes: From nano test tube to nano-reactor, ACS Nano 5, 9306 (2011).

T. Tanaka, H. Liu, S. Fujii, H. Kataura, From metal/semiconductor separation to single-chirality separation of single-wall carbon nanotubes using gel, Phys. Status Solidi-R 5, 301-306 (2011).

J. Wu, K. Gerstandt, H. Zhang, J. Liu and B. J. Hinds, Electrophoretically induced aqueous flow through single-walled carbon nanotube membranes, Nat. Nanotechnol. 7, 133 (2011).

J.-H Han, G. L. C. Paulus, R. Maruyama, D. A. Heller, W.-J. Kim, *et al.* Exciton antennas and concentrators from core-shell and corrugated carbon nanotube filaments of homogeneous composition, Nat. Mater. 9, 833 (2010).

T. Mueller, M. Kinoshita, M. Steiner, V. Perebeinos, A. A. Bol, D. B. Farmer and P. Avouris, Efficient narrow-band light emission from a single carbon nanotube p-n diode, Nat. Nanotechnol. 5, 27 (2009).

H.-J. Shin, S. Clair, Y. Kim and M. Kawai, Substrate-induced array of quantum dots in a single-walled carbon nanotube, Nat. Nanotechnol. 4, 567 (2009).

P. Avouris, M. Freitag and V. Perebeinos, Carbon-nanotube photonics and opto-electronics, Nat. Photon. 2, 341 (2008).

M. Muoth, T. Helbling, L. Durrer, S.-W. Lee, C. Roman, and C. Hierold, Hysteresis-free operation of suspended carbon nanotube transistors, Nat. Nanotechnol. 5, 589 (2010).

R. Singhal, Z. Orynbayeva, R. V. K. Sundaram, J. J. Niu, S. Bhattacharyya, E. A. Vitol, M. G. Schrlau, E. S. Papazoglou, G. Friedman, and Y. Gogotsi, Multifunctional carbon-nanotube cellular endoscopes, Nat. Nanotechnol. 6, 57 (2010).

W. Belzig, Hybrid superconducting devices: Bound in a nanotube, Nat. Phys. 6, 940 (2010).

W. Zhang, Z. Zhang, Y. Zhang, The application of carbon nanotubes in target drug delivery systems for cancer therapies, Nanoscale Res. Lett. 6, 555 (2011).

H. Farhat, S. Berciaud, M. Kalbac, R. Saito, T. F. Heinz, M. S. Dresselhaus, and J. Kong, Observation of electronic Raman scattering in metallic carbon nanotubes, Phys. Rev. Lett. 107, 157401 (2011).

T. Ando, Effects of valley mixing and exchange on excitons in carbon nanotubes with Aharonov-Bohm flux, J. Phys. Soc. Jpn. 75, 024707 (2006).

S. Yasukochi, T. Murai, S. Moritsubo, T. Shimada, S. Chiashi, S. Maruyama, Y. K. Kato, Gate-induced blueshift and quenching of photoluminescence in suspended single-walled carbon nanotubes, Phys. Rev. B 84, 121409(R) (2011).

I. Wilson-Rae, C. Galland, W. Zwerger, and A. Imamoğlu, Nano-optomechanics with localized carbon nanotube excitons, arXiv:0911.1330 (2009).

J. J. Li, W. He and K. D. Zhu, All-optical Kerr modulator based on a carbon nanotube resonator, Phys. Rev. B 83, 115445 (2011).

J. J. Li, C. Jiang, B. Chen and K. D. Zhu, Optical mass sensing with a carbon nanotube resonator, J. Opt. Soc. Am. B (in press) (2012).

J. J. Li and K. D. Zhu, Weighing a single atom using a coupled plasmon-carbon nanotube system, Sci. Technol. Adv. Mat. 13, 025006 (2012).

A. H. Safavi-Naeini, T. P. Mayer Alegre, J. Chan, M. Eichenfield, M. Winger, Q. Lin, J. T. Hill, D. E. Chang, and O. Painter, Electromagnetically induced transparency and slow light with optomechanics, Nature 472, 69 (2011).

V. Fiore, Y. Yang, M. C. Kuzyk, R. Barbour, L. Tian, and H. Wang, Storing optical information as a mechanical excitation in a silica optomechanical resonator, Phys. Rev. Lett. 107, 133601 (2011).

S. Weis, R. Rivière, S. Deléglise, E. Gavartin, O. Arcizet, A. Schliesser, and T. J. Kippenberg, Optomechanically induced transparency, Science 330, 1520 (2010).

J. D. Teufel, D. Li, M. S. Allman, K. Cicak, A. J. Sirois, J. D. Whittaker, and R. W. Simmonds, Circuit cavity electromechanics in the strong-coupling regime, Nature 471, 204 (2011).

R. W. Boyd, Nonlinear optics, (Academic Press, Amsterdam), pp. 313 (2008).

K. F. Graff, Wave motion in elastic solids, (Dover, New York) pp. 539-564 (1991).

C. W. Gardiner, and P. Zoller, Quantum noise, (2nd edn) (Springer, Berlin) pp. 425-433 (2000).

D. F. Walls, and G. J. Milburn, Quantum optics, (Springer, Berlin) pp. 245-265 (1994).

K. L. Ekinci, and M. L. Roukes, Nanoelectromechanical systems, Rev. Sci. Instrum. 76, 061101 (2005).

I. Wilson-Rae, Intrinsic dissipation in nanomechanical resonators due to phonon tunneling, Phys. Rev. B 77, 245418 (2008).

V. Giovannetti, and D. Vitali, Phase-noise measurement in a cavity with a movable mirror undergoing quantum Brownian motion, Phys. Rev. A 63, 023812 (2001).

J. F. Lam, S. R. Forrest and G. L. Tangonan, Optical nonlinearities in crystalline organic multiple quantum wells, Phys. Rev. Lett. 66, 1614 (1991).

A. Lezama, S. Barreiro and A. M. Akulshin, Electromagnetically induced absorption, Phys. Rev. A 59, 4732 (1999).

B. R. Mollow and R. J. Glauber, Quantum theory of parametric amplification, Phys. Rev. 160, 1076 (1967).

R. S. Bennink, R. W. Boyd, C. R. Stroud, and V. Wong, Enhanced self-action effects by electromagnetically induced transparency in the two-level atom, Phys. Rev. A 63, 033804 (2001).

S. E. Harris, J. E. Field, and A. Kasapi, Dispersive properties of electromagnetically induced transparency, Phys. Rev. A 46, R29 (1992).

R. W. Boyd, and D. J. Gauthier, Controlling the velocity of light pulses, Science 326, 1074 (2009).

K. D. Zhu and W. S. Li, Electromagnetically induced transparency due to exciton-phonon interaction in an organic quantum well, J. Phys. B: At. Mol. Opt. Phys. 34, L679 (2001).

A. Subramanian, L. X. Dong, B. J. Nelson, and A. Ferreira, Supermolecular switches based on multiwalled carbon nanotubes, Appl. Phys. Lett. 96, 073116 (2010).

N. Kamaraju, S. Kumar, Y. A. Kim, T. Hayashi, H. Muramatsu, M. Endo, and A. K. Sood, Double walled carbon nanotubes as ultrafast optical switches, Appl. Phys. Lett. 95, 081106 (2009).

F. Zhou, Y. Liu, Z. Y. Li, and Y. Xia, Analytical model for optical bistability in nonlinear metal nano-antennae involving Kerr materials, Opt. Express 18, 13337 (2010).

Z. J. Zhong, Y. Xu, S. Lan, Q. F. Dai, and L. J. Wu, Sharp and asymmetric transmission response in metal-dielectric-metal plasmonic waveguides containing Kerr nonlinear media, Opt. Express 18, 79 (2010).

K. Jacobs, and A. P. Lund, Feedback control of nonlinear quantum systems: a rule of thumb, Phys. Rev. Lett. 99, 020501 (2007).

A. M. Datsyuk, T. Yu. Gromovoi, V. V. Lobanov, Analysis of the properties of carbon nanotubes from distribution maps of molecular electrostatic potential, Theor. & Exp. Chem. 40, 277 (2004).

D. Rugar, R. Budakian, H. J. Mamin, B. W. Chui, Single spin detection by magnetic resonance force microscopy. Nature 430, 329 (2004).

B. Domon, R. Aebersold, Mass spectrometry and protein analysis. Science 312, 212 (2006).

E. Gil-Santos, D. Ramos, A. Jana, M. Calleja, A. Raman, J. Tamayo, Mass sensing based on deterministic and stochastic responses of elastically coupled nanocantilevers. Nano Lett. 9, 4122 (2009).

N. Sinha, J. Z. Ma, J. T. W. Yeow, Carbon nanotube based sensors. Nanosci. J. Nanotechnol. 6, 573 (2006).

K. L. Ekinci, X. M. H. Huang, M. L. Roukes, Ultrasensitive nanoelectromechanical mass detection. Appl. Phys. Lett. 84, 4469 (2004).

Y. T. Yang, C. Callegari, X. L. Feng, K. L. Ekinci, M. L. Roukes, Zeptogram-scale nanomechanical mass sensing. Nano Lett. 6, 583 (2006).

K. Jensen, K. Kim, A. Zettl, An atomic-resolution nanomechanical mass sensor. Nat. Nanotechnol. 3, 533 (2008).

A. K. Naik, M. S. Hanay, W. K. Hiebert, X. L. Feng, M. L. Roukes, Towards single-molecule nanomechanical mass spectrometry. Nat. Nanotechnol. 4, 445 (2009).

B. Lassagne, D. Garcia-Sanchez, A. Aguasca, A. Bachtold, Ultrasensitive mass sensing with a nanotube electromechanical resonator. Nano Lett. 8, 3735 (2008).

K. L. Ekinci, Y. T. Yang, M. L. Roukes, Ultimate limits to inertial mass sensing based upon nanoelectromechanical systems, J. Appl. Phys. 95, 2682 (2004).

V. Sazonova, Y. Yaish, H. Üstünel, D. Roundy, T. A. Arias, P. L. McEuen, A tunable carbon nanotube electromechanical oscillator. Nature 431, 284 (2004).

C. Y. Li, T. W. Chou, Mass detection using carbon nanotube-based nanomechanical resonators. Appl. Phys. Lett. 84, 5246 (2004).

P. A. Greaney, G. Lani, G. Cicero, J. C. Grossman, Anomalous dissipation in single-walled carbon nanotube resonators. Nano Lett. 9, 3699 (2009).

P. Skyba, Notes on measurement methods of mechanical resonators used in low temperature physics, J. Low. Temp. Phys. 160, 219 (2010).

K. C. Schwab, M. L. Roukes, Putting mechanics into quantum mechanics. Phys. Today 58, 36 (2005).

H. Y. Chiu, P. Hung, H. W. Ch. Postma, and M. Bockrath, Atomic-scale mass sensing using carbon nanotube resoantors. Nano. Lett. 8, 4342 (2008).

Z. M. Wang, A. Neogi (eds.), Nanoscale Photonics and Optoelectronics. pp. 27, Springer (2011).

X. Gu, T. Qiu, W. Zhang, P. K Chu, Light-emitting diodes enhanced by localized surface plasmon resonance. Nanoscale Res. Lett. 6, 199 (2011).

A. I. Zhmakin, Enhancement of light extraction from light emitting diodes. Phys. Rep. 498, 189 (2011).

W. Zhang, A. O. Govorov, G. W. Bryant, Semiconductor-metal nanoparticle molecules: Hybrid excitons and the nonlinear Fano effect, Phys. Rev. Lett. 97, 146804 (2006).

W. Zhang and A. O. Govorov, Quantum theory of the nonlinear Fano effect in hybrid metalsemiconductor nanostructures: The case of strong nonlinearity, Phys. Rev. B 84, 081405 (2011).

Z. E. Lu and K. D. Zhu, Enhancing Kerr nonlinearity of a strong coupled exciton-plasmon in hybrid nanocrystal molecules, J. Phys. B 41, 185503 (2008).

Z. E. Lu and K. D. Zhu, Slow light in an artificial hybrid nanocrystal complex, J. Phys. B 42, 015502 (2009).

N. T. Fofang, N. K. Grady, Z. Fan, A. O. Govorov, N. J. Halas, Plexciton Dynamics: Exciton-plasmon coupling in a J-aggregate-Au nanoshell complex provides a mechanism for nonlinearity, Nano Lett. 11, 1556 (2011).

E. Cabrera-Granado, E. Díaz and O. G. Calderón, Slow light in molecular-aggregate nanofilms, Phys. Rev. Lett. 107, 013901 (2011).

S. Kühn, U. Håkanson, L. Rogobete, V. Sandoghdar, Enhancement of single-molecule fluorescence using a gold nanoparticle as an optical nanoantenna. Phys. Rev. Lett. 97, 017402 (2006).

T. Kalkbrenner, U. Håkanson, V. Sandoghdar, Tomographic plasmon spectroscopy of a single gold nanoparticle. Nano Lett. 4, 2309 (2004).

T. Kalkbrenner, U. Håkanson, A. Schädle, S. Burger, C. Henkel, V. Sandoghdar, Optical microscopy via spectral modifications of a nanoantenna, Phys. Rev. Lett. 95, 200801 (2005).

J. Y. Yan, W. Zhang, S. Duan, A. O. Govorov, Optical properties of coupled metal-semiconductor and metal-molecule nanocrystal complexes: Role of multipole effects, Phys. Rev. B 77, 165301 (2008).

A. Yariv, Quantum electronics, (Wiley, New York) (1975).

M. Pelton, J. Aizpurua, G. Bryant, Metal-nanoparticle plasmonics, Laser & Photon. Rev. 2, 136 (2008).

I. C. Khoo, D.H. Werner, X. Liang, A. Diaz, B. Weiner, Nanosphere dispersed liquid crystals for tunable negative-zero-positive index of refraction in the optical and terahertz regimes, Opt. Lett. 31, 2592 (2006).

E. Chang, G. Bussi, A. Ruini and E. Molinari, Excitons in carbon nanotubes: an ab initio symmetry-based approach, Phys. Rev. Lett. 92, 196401 (2004).

R. B. Capaz, C. D. Spataru, S. Ismail-Beigi and S. G. Louie, Diameter and chirality dependence of exciton properties in carbon nanotubes, Phys. Rev. B 74, 121401(R) (2006).

E. Malic, J. Maultzsch, S. Reich and A. Knorr, Excitonic absorption spectra of metallic single-walled carbon nanotubes, Phys. Rev. B 82, 035433 (2010).

Chapter 9

A Circuit Cavity Electromechanical System

Cavity optomechanical systems [Arcizet (2006); Wilson-Rae (2007); Marquardt (2007); Genes (2008); Kippenberg (2008); Groblacher (2009)] and their cavity electromechanical analogues [Woolley (2008); Regal (2008); Teufel (2008); Rocheleau (2010)], usually composed of a mechanical resonator coupled with a microwave or optical cavity via radiation pressure force, have been under extensive investigation. For example, the Cooper pair boxes [Armour (2002); Wallraff (2004)] or superconducting microwave cavity [Regal (2008)] coupled to nanomechanical resonator (NR). Particularly, the coupled nanomechanical resonator-superducting microwave cavity system has been used to investigate quantum entanglement, nanomechanical squeezing [Vitali (2007); Woolley (2008); Tian (2008)] and back-action evading measurement [Hertzberg (2010)]. More recently, cooling the nanomechanical resonator based on this system has been proposed theoretically [Xue (2007); Li (2008); Teufel and Regal (2008)] and realized experimentally [Rocheleau (2010)]. For example, Rocheleau et al. [Rocheleau (2010)] have reported the cooling of the motion of a radio-frequency nanomechanical resonator by parametric coupling to a driven, microwave-frequency superconducting resonator. Starting from a thermal occupation of 480 quanta, they have observed occupation factors as low as 3.8 ± 1.3 and found the mechanical resonator with probability 0.21 in the quantum ground state of motion.

However, in cavity optomechanical systems, the mechanical interactions between the resonator and the cavity modes can modify the optical response, which will eventually results in the effects such as parametric normal-mode splitting and optomechanically induced transparency (OMIT). Schliesser et al. have realized optomechanically induced transparency (OMIT) [Schliesser (2009)] in optomechanical systems, which is in analogy with the effect of EIT in atomic system. In particular, experimental realization of EIT effect in electromechanical system [Teufel (2008)] make it a promising candidate for photonic devices in the microwave regime. In this chapter, we study a driven nanomechanical resonator capacitively coupled to a superconducting microwave cavity, in the presence of two optical microwaves. Due to the investigation of the coherent optical spectrum of this system, we propose a feasible way to realize a single photon router, a tunable four-wave mixing process [Jiang (2012)], and a microwave mass sensor [Jiang (2011)], which may have a variety of potential applications such as quantum transducer and quantum information processing.

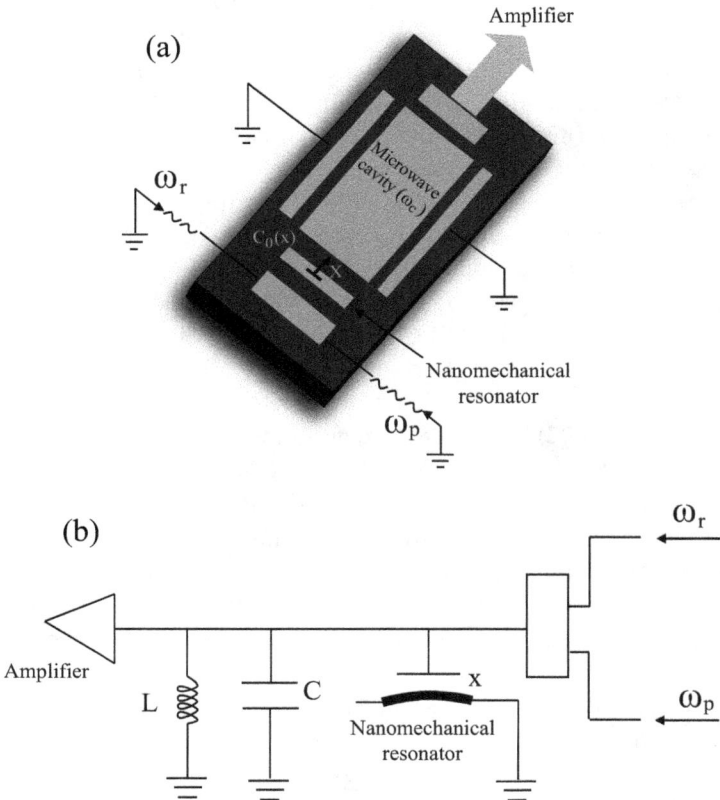

Fig. 9.1 Schematic of a nanomechanical resonator capacitively coupled to a microwave cavity denoted by equivalent inductance L and equivalent capacitance C in the presence of a strong pump field ω_p and a weak probe field ω_r. (b) Equivalent circuit.

The cavity electromechanical system is plotted in Fig.9.1, where a nanomechanical resonator with resonance frequency ω_n is capacitively coupled to a superconducting microwave cavity denoted by the equivalent inductance L and equivalent capacitance C. In the presence of a strong pump field with frequency ω_p and a weak probe field with frequency ω_r, the radiation pressure force oscillating at the beat frequency between the pump field and the probe field can produce the vibration of nanomechanical resonator, which is the same with typical cavity optomechanical system. As usual, the vibration modes of nanomechanical resonator can be treated as phonon modes.

The vibration of nanomechanical resonator will induce the displacement x from its equilibrium position, which will in turn alters the capacitance of the microwave cavity. Here, the coupling capacitance can be approximated by $C_0(x) = C_0(1 - x/d)$, where C_0 and d represent an equilibrium capacitance and the equilibrium nanoresonator-cavity separation, respectively. In this case, the coupled cavity has an equivalent capacitance $C_\Sigma = C + C_0(x)$, which corresponds to the resonance frequency of the microwave cavity is $\omega_c = 1/\sqrt{LC_\Sigma}$.

The interaction Hamiltonian between the mechanical resonator and the microwave cavity is $H_I = \hbar g_0 a^\dagger a(b^\dagger + b)$, where $a^\dagger$ (a) and $b^\dagger$ (b) are the creation (annihilation) operators for photons and phonons, respectively. The single-photon coupling rate $g_0 = g\delta x_{zp}$ is the product of $g = \partial\omega_c/\partial x$ and the zero point motion $x_{zp} = \sqrt{\hbar/2m_{\text{eff}}\omega_n}$ of the mechanical resonator, where m_{eff} and ω_n are the effective mass and vibrational frequency of NR, respectively.

In a rotating frame at the pump frequency ω_p, the Hamiltonian of the coupled nanomechanical resonator and a superconducting microwave cavity can be expressed as [Regal (2008)]

$$
\begin{aligned}
H = {} & \hbar\Delta_p a^\dagger a + \hbar\omega_n b^\dagger b - \hbar g_0 a^\dagger a(b^\dagger + b) \\
& + i\hbar(E_p a^\dagger + E_p^* a) + i\hbar(E_r a^\dagger e^{-i\delta t} + E_r^* a e^{i\delta t}),
\end{aligned}
\tag{9.1}
$$

where ω_c is the resonance frequency of microwave cavity. $\Delta_p = \omega_c - \omega_p$ and $\delta = \omega_r - \omega_p$ denote the cavity-pump detuning and probe-pump detuning, respectively. The last two terms represent the interaction between the cavity field and the two microwave fields. E_p and E_r are, respectively, the amplitudes of pump field and probe field, and they are defined by

$$
|E_p| = \sqrt{2P_p\kappa/\hbar\omega_p},
\tag{9.2}
$$

$$|E_r| = \sqrt{2P_r\kappa/\hbar\omega_r}, \tag{9.3}$$

where P_p and P_r are the pump power and probe power, respectively. κ represent the decay rate of the cavity. Here, in the presence of a strong pump field, the probe field can be treated as the mean response of the coupled system, which made the quantum fluctuations neglected during the theoretical calculation [Weis (2010)]. The same treatment can also be found in the context of EIT where one uses atomic mean value equations and all quantum fluctuations due to both spontaneous emission and collisions are neglected [Hau (1999)].

The Heisenberg equation of motion corresponding to Hamiltonian (9.1) can be written as

$$\frac{d\langle a \rangle}{dt} = -(i\Delta_p + \kappa)\langle a \rangle + ig_0 \langle a \rangle \langle Q \rangle + E_p + E_r e^{-i\delta t}, \tag{9.4}$$

$$\frac{d^2\langle Q \rangle}{dt^2} + \gamma_n \frac{d\langle Q \rangle}{dt} + \omega_n^2 \langle Q \rangle = 2\omega_n g_0 \langle a^\dagger \rangle \langle a \rangle, \tag{9.5}$$

where $Q = b^\dagger + b$ is the phonon amplitude of the resonator, and $\langle a \rangle$, $\langle a^\dagger \rangle$, $\langle Q \rangle$ are the expectation values of the operators a, $a^\dagger$, and Q, respectively [Greence (1988)]. γ_n is the damping rate of the mechanical mode. To solve the above equations, we make the ansatz [Boyd (2008)] $\langle a(t) \rangle = a_0 + a_+ e^{-i\delta t} + a_- e^{i\delta t}$, and $\langle Q(t) \rangle = Q_0 + Q_+ e^{-i\delta t} + Q_- e^{i\delta t}$. Since the probe field E_r is much weaker than the pump field E_p, upon substituting these equations into Eqs.(9.4) and (9.5) and working to the lowest order in E_r but to all orders in E_p, we obtain in the steady state:

$$a_+ = \frac{\kappa + \theta - i(\delta + \Delta_p)}{(\kappa - i\delta)^2 - (\theta - i\Delta_p)^2 - \beta} E_r, \tag{9.6}$$

$$a_- = \frac{i\alpha\eta^*\omega_n}{(\kappa + i\Delta_p - i\alpha\omega_n n_p)^2} \cdot \frac{E_p^2 E_r}{(\kappa + i\delta)^2 - (\theta^* + i\Delta_p)^2 - \beta^*}, \tag{9.7}$$

where $\eta = \omega_n^2/(\omega_n^2 - \delta^2 - i\gamma_n\delta)$, $\alpha = 2g_0^2/\omega_n^2$, $\beta = \alpha^2\eta^2\omega_n^2 n_p^2$, $\theta = i\alpha\omega_n n_p(\eta + 1)$, and intracavity photon number $n_p = |a_0|^2$ is determined by the following equation

$$n_p \left[\kappa^2 + (\Delta_p - \omega_n\alpha n_p)^2\right] = |E_p|^2. \tag{9.8}$$

The output field transmitted through the coupled nanomechanical res-onator and a superconducting microwave cavity can be obtained using the standard input-output theory [Gardiner (2004)] $a_{\text{out}}(t) = a_{\text{in}}(t) - \sqrt{2\kappa}a(t)$, where $a_{\text{out}}(t)$ is the output field operator. We have

$$
\begin{aligned}
\langle a_{\text{out}}(t) \rangle &= a_{\text{out}0} + a_{\text{out}+}e^{-i\delta t} + a_{\text{out-}}e^{i\delta t} = \sqrt{2\kappa}(a_0 + a_+ e^{-i\delta t} + a_- e^{i\delta t}) \\
&= (E_p - \sqrt{2\kappa}a_0)e^{-i\omega_p t} + (E_r - \sqrt{2\kappa}a_+)e^{-i(\delta+\omega_p)t} \\
&\quad - \sqrt{2\kappa}a_- e^{i(\delta-\omega_p)t} \\
&= (E_p - \sqrt{2\kappa}a_0)e^{-i\omega_p t} + (E_r - \sqrt{2\kappa}a_+)e^{-i\omega_r t} \\
&\quad - \sqrt{2\kappa}a_- e^{-i(2\omega_p-\omega_r)t}.
\end{aligned}
\tag{9.9}
$$

Here, the output field contains two input components (ω_p and ω_r) and one new component $2\omega_p - \omega_r$. The transmission of the probe field, defined by the ratio of the output and input field amplitudes at the probe frequency, can be expressed as

$$
\begin{aligned}
T &= \frac{E_r - \sqrt{2\kappa}a_+}{E_r} \\
&= 1 - \sqrt{2\kappa}\frac{\kappa + \theta - i(\delta + \Delta_p)}{(\kappa - i\delta)^2 - (\theta - i\Delta_p)^2 - \beta}.
\end{aligned}
\tag{9.10}
$$

Likewise, the nonlinear part of Eq.(9.9) with frequency $2\omega_p - \omega_r$ corresponds to the nonlinear four-wave mixing (FWM) process, which is a new generation term and can be defined as

$$
\begin{aligned}
\text{FWM} &= \left| \frac{\sqrt{2\kappa}a_-}{E_r} \right|^2 \\
&= \left| \frac{\sqrt{2\kappa}i\alpha\eta^*\omega_n}{(\kappa + i\Delta_p - i\alpha\omega_n n_p)^2} \cdot \frac{E_p^2}{(\kappa + i\delta)^2 - (\theta^* + i\Delta_p)^2 - \beta^*} \right|^2.
\end{aligned}
\tag{9.11}
$$

The FWM conversion efficiency η_{FWM} is given by [Huang (2010); Li (2011)]

$$
\eta_{\text{FWM}} = \frac{P_i}{P_r} = \frac{\hbar(2\omega_p - \omega_r)|\sqrt{2\kappa}a_-|^2}{P_r} = \left| \frac{2\kappa a_-}{E_r} \right|^2,
\tag{9.12}
$$

where P_i is the power of the generated idler signal.

We choose the parameters of the experimentally realized cavity electromechanical system as follows [Rocheleau (2010)]: $(\omega_c, \omega_n, \kappa, g_0) = (2\pi \times 7.5\text{GHz}, 2\pi \times 6.3\text{MHz}, 2\pi \times 600\text{kHz}, 250\text{Hz})$. $Q_n = 10^6$ is the quality factor of the nanomechanical resonator and related to the damping rate γ_n given by ω_n/Q_n. The system works in the resolved-sideband regime ($\omega_n > \kappa$) termed good-cavity limit, a prerequisite for resolved-sideband cooling of the micromechanical oscillator [Schliesser (2009)].

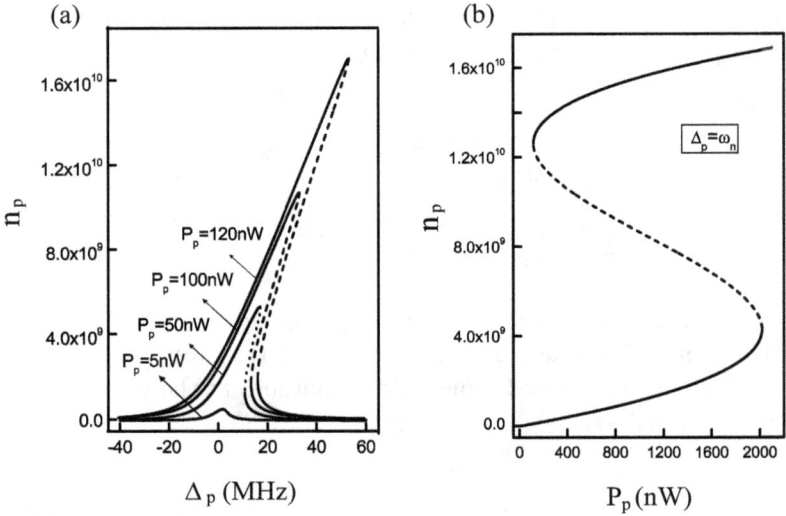

Fig. 9.2 Steady-state intracavity photon number as a function of (a) cavity-pump detuning Δ_p for $P_p = 5, 50, 100$, and 120nW; (b) pump power for $\Delta_p = \omega_n$. Other parameters used are $\omega_c = 2\pi \times 7.5$GHz, $\omega_n = 2\pi \times 6.3$MHz, $\kappa = 2\pi \times 600$kHz, $g_0 = 2\pi \times 460$Hz, and $\gamma_n = 2\pi \times 30$Hz. The dashed curves and solid curves correspond to the unstable states and stable states, respectively.

9.1 Coherent optical spectrum

Firstly, we investigate the optomechanical bistability in this coupled system. The form of cubic Eq.(9.8) is characteristic of optical multistability [Gupta (2007); Brennecke (2008); Kanamoto (2010)]. Figure 9.2(a) shows the mean intracavity photon number as a function of the cavity-pump detuning Δ_p with different pump powers. Clearly, when the pump power equals to 5nW, the curve is nearly Lorentzian. Increasing the pump power, the bistable behavior of the coupled nanomechanical resonator and a superconducting microwave cavity becomes more obvious. That is to say, when increasing the pump power the initially Lorentzian resonance curve gets asymmetric and develops an increasing region with three possible states above a critical value. Figure 9.2(a) further shows that the largest and smallest roots of Eq.(9.8) correspond to the stable states, while the middle root of Eq.(9.8) is unstable state. When the cavity is driven on its red sideband ($\Delta_p = \omega_n$), Fig.9.2(b) shows an appropriate regime for cooling the mechanical resonator close to the quantum ground state [Teufel (2008)].

Optomechanical bistability provides a candidate for some controlled switching devices, since its intracavity photon number for the lower stable branch and the upper stable branch can be simply controlled by the input pump power and the cavity-pump detuning. There are two differences for optical bistability in atomic system and optomechanical system: (1) Optical bistability in atomic systems refers to the bistability in the input-output intensity while optomechanical bistability refers to the bistable behavior of the intracavity photon number. (2) The nonlinearity in atomic systems occurs in its polarization response (i.e., it arises from the internal degrees of freedom of atoms) [Venkatesh (2011)], while the optomechanical bistability is the result of a competition between mechanical restoring force and radiation pressure force [Ghobadi (2011)].

Here, we mainly consider the situation where the cavity is driven on its blue sideband, i.e., $\Delta_p = -\omega_n$. Under blue-detuned pumping, the effective interaction Hamiltonian for the cavity field and the mechanical phonon mode relates to the case of parametric amplification, $H_I = \hbar G(a^\dagger b^\dagger + ab)$, where $G = g_0\sqrt{n_p}$ is the effective coupling strength. Figure 9.3 displays a series of transmission spectra of the probe field as a function of the probe-cavity detuning ($\Delta_r = \omega_r - \omega_c$) for various pump powers. Figure 9.3(a) shows the transmission spectrum of the probe field in the absence of the pump field, which is the usual Lorentzian line shape of the bare cavity. However, as the pump power is raised, we can see from Fig.9.3(b) and Fig.9.3(c) that the transmission is attenuated around the probe-cavity detuning $\Delta_r = 0$ compared to that in Fig.9.3(a), a result of the increased feeding of photons into the cavity. If the pump power is increased further, the system switches from electromagnetically induced absorption (EIA) [Lezama (1999)] to parametric amplification (PA) [Mollow (1967)], leading to probe amplification (Fig.9.3(d)-(f)). The switching point is $C \approx 1$. The physical origin of this phenomenon comes from the radiation pressure force oscillating at the beat frequency between the pump field and the probe field, which induces the vibration of the nanomechanical resonator. When the beat frequency is resonant with the mechanical resonance frequency ω_n. The frequency of the pump field ω_p is downshifted to the Stokes frequency $\omega_p - \omega_n$, which is degenerate with the probe field. Constructive interference between the Stokes field and the probe field amplifies the weak probe field. Similar amplification of a probe due to radiation pressure backaction in a detuned cavity optomechanical system was recently demonstrated by Verlot *et al.* [Verlot (2010)]. Note that the phenomenon of parametric oscillation instability can occur at pump power threshold when the the Stokes field

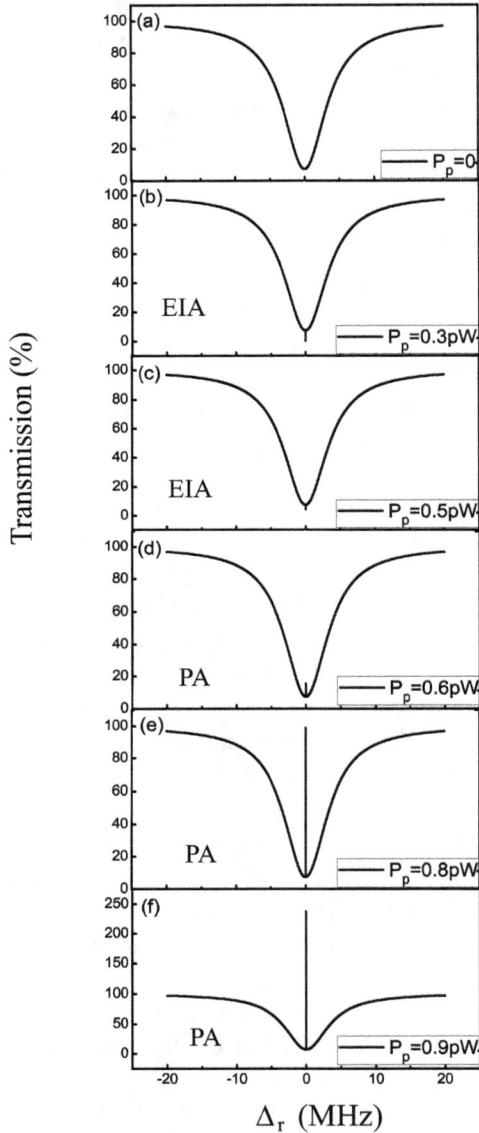

Fig. 9.3   The normalized magnitude of the cavity transmission $|T|^2$ as a function of probe-cavity detuning $\Delta_r = \omega_r - \omega_c$ for various pump powers. Pump powers (pW) are: 0 (a), 0.3 (b), 0.5 (c), 0.6 (d), 0.8 (e), 0.9 (f), respectively. When the pump power increases, the system switches from EIA to PA. Other parameters used are $\Delta_p = -\omega_n$, $\omega_c = 2\pi \times 7.5$GHz, $\kappa = 2\pi \times 600$kHz, $g_0 = 2\pi \times 460$Hz, $\gamma_n = 40$Hz, and $\omega_n = 2\pi \times 6.3$MHz.

coincides with the cavity resonance, which has been predicted by Braginsky [Braginsky (2001)] and demonstrated for the first time at Caltech university [Weis (2010)].

Figure 9.4 shows the probe absorption spectrum as a function of the the probe-pump detuning δ. Here, we note that two sideband peaks appear exactly at $\delta = \pm\omega_n$ of the probe absorption spectrum. These two sharp peaks demonstrate the resonant absorption and amplification of the mechanical mode. An intuitive physical picture explaining these peaks can be given in the dressed-states shown in Fig.9.4(b). The uncoupled energy levels on the left side split into dressed states $|n, m\rangle$ shown in Part (1). Part (2) describes a process in which the photon makes a transition from the lowest dressed level $|n - 1, m\rangle$ to the highest level $|n, m + 1\rangle$ by the simultaneous absorption of two pump photons and emission of a photon at

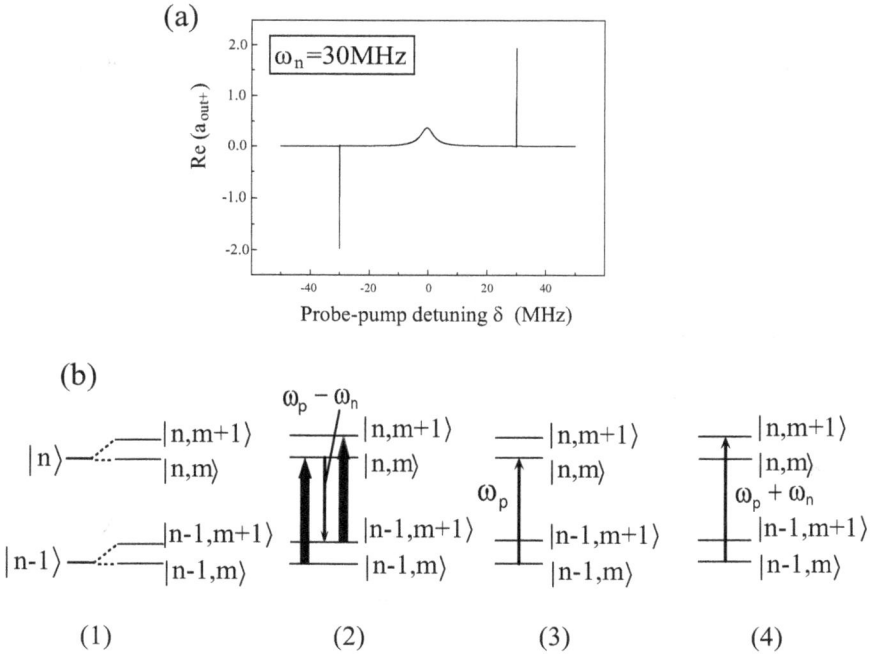

Fig. 9.4   (a) The plot of $Re(a_{\mathrm{out}+})$ as a function of probe-pump detuning δ with $\omega_n = 30$MHz. Other parameters are $P_p = 0.3$nW, $\omega_c = 2\pi \times 7.5$GHz, $\kappa = 2\pi \times 600$kHz, $g_0 = 250$Hz, $\gamma_n = 2\pi \times 30$Hz and $\Delta_p = 0$. (b) The transitions between the dressed states corresponding to the peaks shown in (a), where $|n\rangle$ and $|m\rangle$ denote the cavity mode and phonon mode, respectively.

frequency $\omega_p - \omega_n$, corresponding to the negative absorption (amplification) at $\delta = -\omega_n$ in Fig.9.4(a). Part (3) in Fig.9.4(b) shows the usual absorption of the cavity. And Part (4) corresponds to the absorption at frequency $\delta = \omega_n$. Therefore, Fig.9.4 provides a simple method to measure the resonance frequency of the nanomechanical resonator. If we fix cavity-pump field detuning $\Delta_p = 0$ and scan the probe frequency across the microwave cavity frequency, then we can easily obtain the resonance frequency of the nanomechanical resonator in the probe absorption spectrum, due to the steep sideband peaks.

9.2   Single-photon router with a cavity electromechanical system

In the past decade, a single-photon router or an optical switch has been under extensive exploration, which provides an potential for quantum node in the quantum information networks [Kimble (2008)]. Some remarkable works include: Aoki *et al.* have realized a single-photon quantum router with high efficiency for photon storing by one atom and a microtoroidal cavity [Aoki (2009)]. Based on electromagnetically induced transparency (EIT), Bajcsy *et al.* [Fleischhauer (2005)] demonstrated an efficient optical switch by confining both photons and a small laser-cooled atomic ensemble of atoms inside a hollow fiber [Bajcsy (2009)]. Recently, Hall *et al.* [Hall (2011)] reported ultrafast switching of photonic entanglement and proposed its use as a single-photon router. In 2012, Agarwal and Huang theoretically demonstrated the possibility of using optomechancial systems as a single-photon router [Agarwal (2012)]. Despite of the success of laser-based control, it is desirable to develop a microwave approach due to its relatively easy generation and control. Very recently, scattering of incident microwave by a single artificial atom coupled to an open transmission line has been studied and EIT effect was observed [Astafiev (2010); Abdumalikov (2010)]. After that, Hoi *et al.* employed EIT in this system to demonstrate a single-photon router by measuring simultaneously the transmission and reflection of the incident microwave photons [Hoi (2011)].

In this section, we shall present a scheme for implementing a single-photon router in a cavity electromechanical system which operates in the microwave regime. With and without the pump field, the probe photons are transmitted and reflected, due to optomechanically induced transparency (OMIT). Therefore, we can apply a tunable pump field to switch the route of the probe field. Compared with [Agarwal (2012)], we need a lower pump power of dozens of picowatt, on the same order of magnitude with Ref.[Hoi (2011)], to realize the switch. In addition, on-chip operation of the electromechanical system enables the router to be scalable, which is more applicable in real quantum information networks.

We consider the cavity electromechanical system, shown schematically in Fig.9.5 (the dashed rectangle), where a nanomechanical resonator is capacitively coupled to a superconduting microwave cavity denoted by equivalent inductance L and capacitance C. A strong pump microwave field at the frequency ω_p is applied through the coaxial lines to route a weak microwave signal at the probe frequency ω_r. The transmitted and reflected probe field can be measured simultaneously via a low noise, cryogenic amplifier (triangle). The circulators, number $1-4$, enable us to separate signals propagating in different directions in the lines [Hoi (2011)]. The displacement x of the nanomechanical resonator from its equilibrium position alters

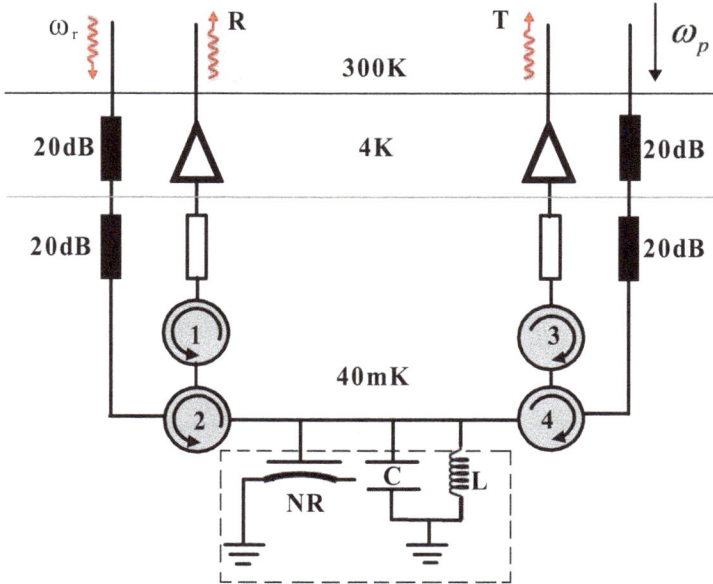

Fig. 9.5   Schematic of the measurement setup. A strong pump field at the frequency ω_p and a weak probe field at the frequency ω_r are applied through coaxial lines, which inductively couple to the microwave cavity. A nanomechanical resonator (NR) is capacitively coupled to the cavity denoted by equivalent inductance L and capacitance C (the dashed rectangle). The transmission and reflection of the probe field can be measured simultaneously via the circulators numbered 1-4.

the capacitance C of the microwave cavity and thus its resonance frequency Ω_c. The system operates in the resolved-sideband regime is beneficial for ground state cooling of mechanical motion [Teufel and Donner (2011)]. In what follows, we consider the situation that the microwave cavity is driven on its red sideband, i.e., $\Delta_c = \omega_n$, due to the optomechanically induced transparency.

Figure 9.6(a) plots the magnitude of the probe transmission and reflection as a function of the probe-cavity detuning when the pump field is off. In the absence of the pump field, the probe field is totally reflected on

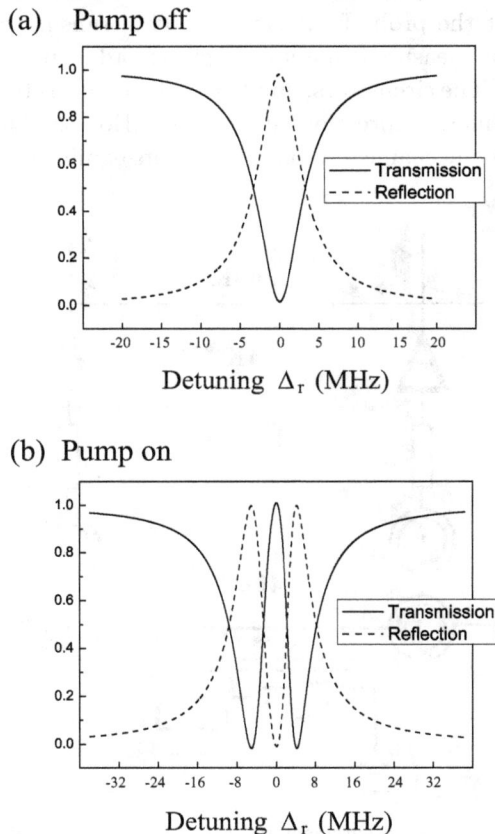

(a)   **Pump off**

(b)   **Pump on**

Detuning Δ_r (MHz)

Fig. 9.6   The magnitude of transmission and reflection as a function of probe-cavity detuning (a) in the absence of the pump field, (b) the pump power equals to 20nW and $\Delta_p = \omega_n$. The parameters used are $\omega_c = 2\pi \times 7.5\text{GHz}$, $\omega_n = 2\pi \times 10.69\text{MHz}$, $\kappa = 2\pi \times 170\text{kHz}$, $\gamma_n = 2\pi \times 30\text{Hz}$, $Q_n = 360{,}000$, and $g_0 = \pi \times 460\text{Hz}$.

resonance and the transmission is zero. However, when the pump field turns on, this situation is totally changed. As shown in Fig.9.6(b), when the pump power P_p equals to 20nW, the phenomenon of OMIT based on the normal-mode splitting (NMS) appears. The original Lorentzian dip of the probe transmission splits into a doublet with a separation $2g_0$, and the cavity electromechanical system becomes transparent to the probe field at $\Delta_r = 0$. The physical origin of this phenomenon can be understood as a result of a radiation pressure force at the beat frequency between the pump and probe frequency, which drives the motion of the nanomechanical resonator near its resonance frequency. This in turn gives rise to Stokes $(\omega_s = \omega_c - \omega_n)$ and anti-Stokes $(\omega_{as} = \omega_c + \omega_n)$ scattering of photons from the strong intracavity field. The Stokes scattering is strongly suppressed because it is highly off-resonant with the microwave cavity, while the anti-Stokes field interfering with near-resonant probe field and eventually modify the probe spectrum.

In view of the above analysis, we exploit OMIT to create a single-photon router. Figure 9.7 demonstrates the operation principle. Applying two

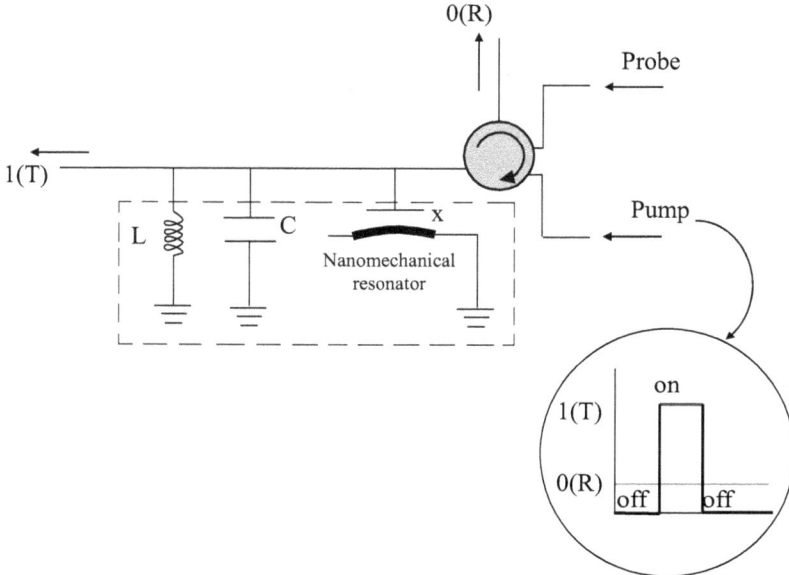

Fig. 9.7 Cartoon of the router. When the pump field is off, the probe field is reflected to the port 0(R); when the pump filed is on, the probe field is transmitted to the port 1(T). The inset denote the pump pulse.

optical fields on the cavity electromechanical system, one is a weak, continuous probe field at the frequency $\omega_p = \omega_c$, and the other is a strong, tunable pump field with detuning $\Delta_p = \omega_m$. There are two situations: (1)when the pump field is off, the photons are reflected by the cavity electromechanical system and travel through the circulator to the output 0; (2)when the pump field turns on, the photons are transmitted and travel to output 1, due to OMIT. Our routing principle is similar to that of a recently realized single-photon router in the microwave regime, where a transmon qubit is embedded in an open transmission line [Hoi (2011)].

Figure 9.8(a) plot the magnitude of transmission as a function of probe-cavity detuning with different pump powers. Clearly, the transmission at $\Delta_r = 0$ increases with the stronger pump power, i.e., when $P_p = 50\text{pW}$, the magnitude of transmission approximately equals to 1. The six curves in Fig.9.8(a) show that the transparency window will be broadened as the

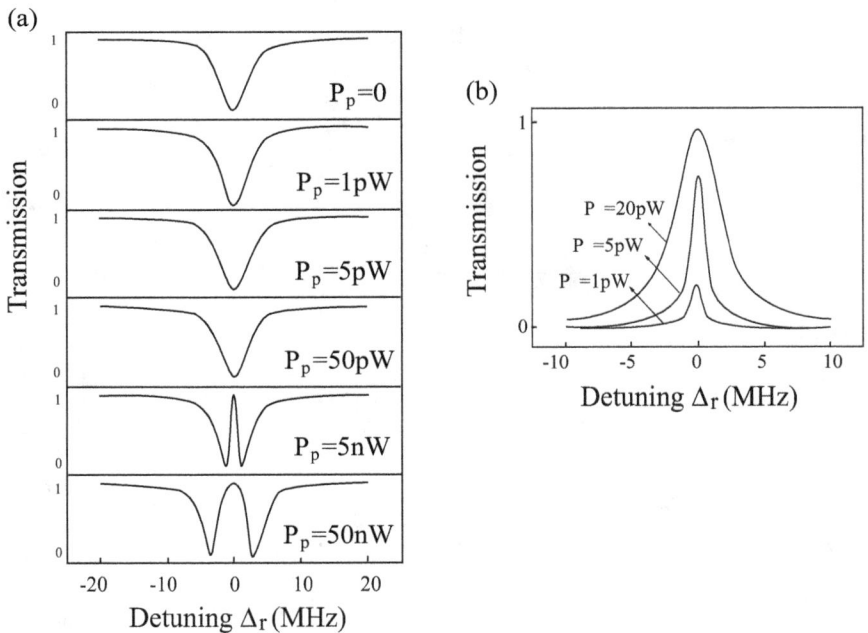

Fig. 9.8   (a) The magnitude of probe transmission as a function of the probe-cavity detuing for pump power P_p equals to 0, 1pW, 5pW, 50pW, 5nW, and 50nW, respectively. (b) The enlarged images of the central peak in (a) for $P_p = 1\text{pW}$, 5pW, and 20pW, which shows clearly that the transmission increases with larger pump power. The parameters used are the same with Fig.9.6.

pump power increases. Figure 9.8(b) is the enlarged images of the narrow peaks in the center of Fig.9.8(a) for three different pump powers.

Figure 9.9 plots the magnitude of transmission and reflection as a function of pump power when the probe field is resonant with the cavity field. It shows that when the pump field is off, the photons are totally reflected ($|R|^2 \approx 1$ and $|T|^2 \approx 0$). When the pump power turns on, the magnitude of reflection decreases while the transmission increases. For intermediate values of pump power, a portion of power, quantified by $1 - |T|^2 - |R|^2$, is lost due to the decay of the cavity. However, we notice that the probe field can be transmitted totally ($|T|^2 \approx 1$ and $|R|^2 \approx 0$) when the pump power reach 50pW. Such low pump power is enough to convert the output probe field from reflection to transmission, which enables the cavity electromechanical system to behave as a single-photon router in the microwave regime. Furthermore, on-chip operation of the electromechanical system enables the operation scheme of the router to be scalable.

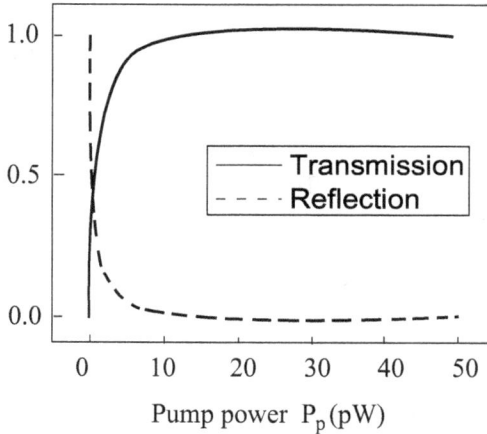

Fig. 9.9   The magnitude of transmission and reflection as a function of the pump power. The parameters used are the same with Fig.9.6.

9.3   Controllable nonlinear responses

Since the advent of nonlinear optics, four-wave mixing (FWM) — an important third-order nonlinear process has received a lot of research interest, which makes great contributions in all-optical communication networks and all-optical switching [Diez (1997)]. Furthermore, because of the excellent amplitude, stable frequency and narrow bandwidth, microwave photonic system is the best candidate for the generation of four-wave mixing process [Kitayama (1997)]. Recently, by using optical four-wave mixing, Wiberg *et al.* have achieved the microwave-photonic frequency multiplication [Wiberg (2006)]. However, in conventional materials, the nonlinear FWM coefficient can be affected by the linear absorption loss, which will decrease the magnitude of nonlinear process. In this case, Harris *et al.* have proposed an efficient way to resonantly enhance the nonlinear process using electromagnetically induced transparency (EIT) in a three-level system [Harris (1990); Boller (1991)]. Since this pioneering work, various schemes were presented to enhance the nonlinear process based on EIT in both atomic systems and solid state systems. For example, Li and Xiao have observed the enhancement of nondegenerate four-wave mixing (NDFWM) based on EIT in a lambda-type three-level system of rubidium atoms [Li (1996)]; Schmidt and Imamoğlu have proposed a scheme to obtain giant Kerr nonlinearities with vanishing linear susceptibilities by EIT in a four-level atomic system [Schmidt (1996)]; Ham *et al.* have observed EIT in an inhomogeneously broaden spectral hole-burning system of Pr:YSO and showed the enhancement of FWM with the reduction of absorption [Ham (1997)].

Furthermore, the quantum-enabled strong coupling regime and the sideband cooling of the micromechanical motion of cavity electromechanical system have been investigated by Teufel' group [Teufel and Li (2011); Teufel and Donner (2011)]. In this section, we study the nonlinear four-wave mixing process in this system, where the cavity is driven by a strong pump field at frequency ω_p and a weak probe field at frequency ω_r. Due to the radiation pressure, two pump photons would mix with a probe photon to yield an idler photon at frequency $\omega_i = 2\omega_p - \omega_r$. Theoretical study shows that the generated FWM intensity can be greatly enhanced without linear absorption, due to OMIT effect. Increasing the pump power, the FWM intensity will be enhanced gradually. It should be pointed out that, in the microwave regime the nonlinear FWM process can be realized with many different techniques and may be purchased commercially, while in optical domain, such FWM is hard to recognize. In this section, we give more physical insight to the resonantly enhanced FWM in the cavity electromechanical system which is compatible with on chip operation, and we obtain a high FWM conversion efficiency at an extremely low pump power.

The cavity electromechanical system achieving four-wave fixing process is sketched in Fig.9.1(a), where a nanomechanical resonator with resonance frequency ω_n is capacitively coupled to a superconducting microwave cavity denoted by the equivalent inductance L and equivalent capacitance C. When the cavity is driven by a strong pump field at frequency ω_p and a weak probe field at frequency ω_r simultaneously, the radiation pressure force oscillating at the beat frequency between the pump field and the probe field induces the nanomechanical resonator to vibrate.

(a)

(b)

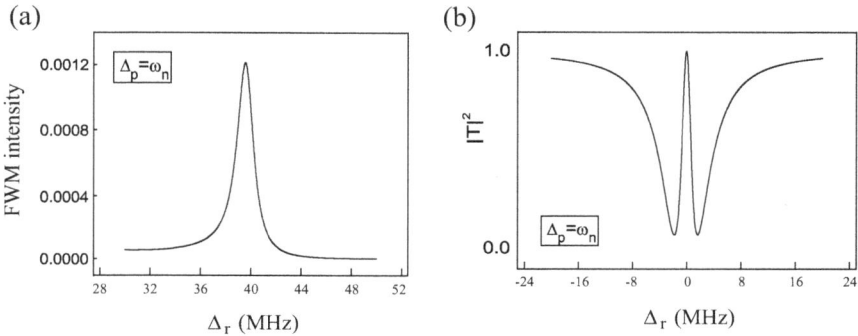

Fig. 9.10 (a) FWM intensity as a function of cavity-pump detuning with probe-cavity detuning Δ_r. (b) The normalized magnitude of the cavity transmission as a function of the Δ_r. The parameters are $P_p = 50\text{nW}$, $\omega_c = 2\pi \times 7.5\text{GHz}$, $\Delta_p = \omega_n = 2\pi \times 6.3\text{MHz}$, $\kappa = 2\pi \times 600\text{kHz}$, $g_0 = \pi \times 460\text{Hz}$, and $\gamma_n = 2\pi \times 30\text{Hz}$.

Figure 9.10(a) plots the shape of the FWM spectrum as a function of the cavity-pump detuning for $P_p = 50\text{nW}$ when the probe field is resonant with the cavity field. It can be seen that there is a peak at $\Delta_r = \omega_n$, which occurs in such cavity opto- or electro-mechanical system when the cavity is driven on the red sideband ($\Delta_p = \omega_n$). This effect is attributed to OMIT, which has been extensively investigated by [Agarwal (2010); Weis (2010); Safavi-Naeini (2011); Teufel and Li (2011)]. Figure 9.10(b) shows the linear transmission spectrum of the probe field with $P_p = 50\text{nW}$, where a transparency window is appeared at $\Delta_r = 0$. When the pump power increases, the transparency window will be broadened. Therefore, the FWM process can be resonantly enhanced with vanshied linear absorption based on OMIT in cavity electromechanical system [Li (1996)].

Figure 9.11 plots the shape of the FWM spectrum as a function of the probe-cavity detuning when the cavity is driven on the red sideband, i.e., $\Delta_p = \omega_n$. Figure 9.11(a) shows that when the pump power increases, the FWM intensity near the resonant region is enhanced significantly, and the linewidth of the spectrum is broadened simultaneously. When the pump power increases, the radiation pressure force and the effective coupling

(a)

(b)

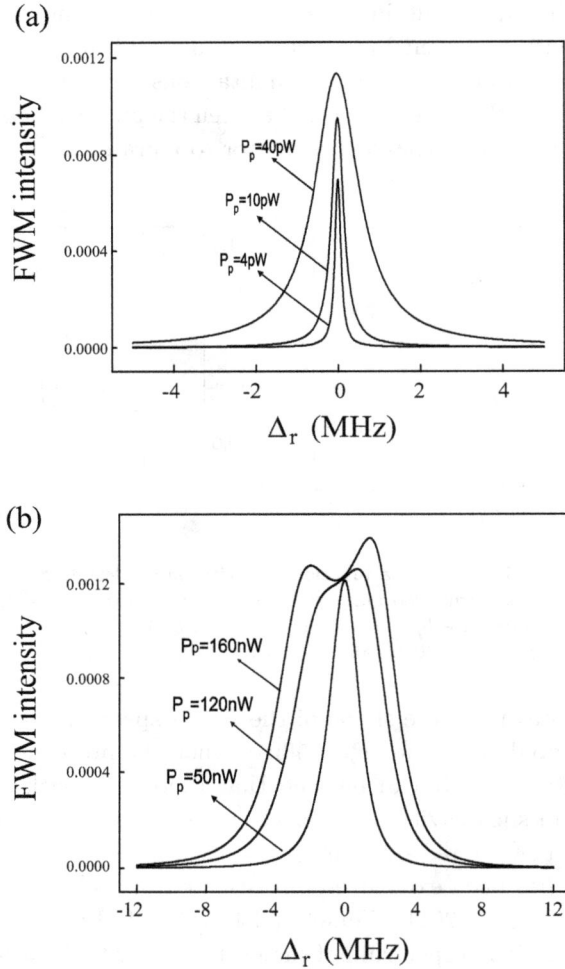

Fig. 9.11   FWM intensity versus probe-cavity detuning with $\Delta_p = \omega_n$ for (a) $P_p = 4, 10,$ and 40pW; (b) $P_p = 50, 120,$ and 160nW. The parameters used are $\omega_c = 2\pi \times 7.5\text{GHz}$, $\omega_n = 2\pi \times 6.3\text{MHz}$, $\kappa = 2\pi \times 600\text{kHz}$, $g_0 = \pi \times 460\text{Hz}$, and $\gamma_n = 2\pi \times 30\text{Hz}$.

strength between the cavity and the resonator will be enhanced, due to the more intracavity photon numbers. On the other hand, in the absence of the interaction between the cavity field and the resonator ($g_0 = 0$), no FWM component appears in the output field, according to Eq.(9.11). Therefore, we can conclude that the parametric coupling in the system plays a vital role in generating FWM, and the FWM intensity increases when the cavity is driven by a stronger pump field.

The magnitude of the mixture can be enhanced by increasing the pump power. Although the resonantly enhanced FWM occurs when the probe-cavity detuning is very small, on the order of kHz, it is still possible to detect the FWM component in real experiments using the cryogenic low-noise amplifier [Teufel and Li (2011)]. Furthermore, Fig.9.11(b) displays that increasing the pump power above a critical value can make the FWM spectrum evolve from one peak to double peaks, while the maximum value in the resonant region remains almost constant. The critical value of the transition from one peak to double ones is about 120nW, which almost coincides with the value where the bistable behavior occurs in Fig.9.2(b).

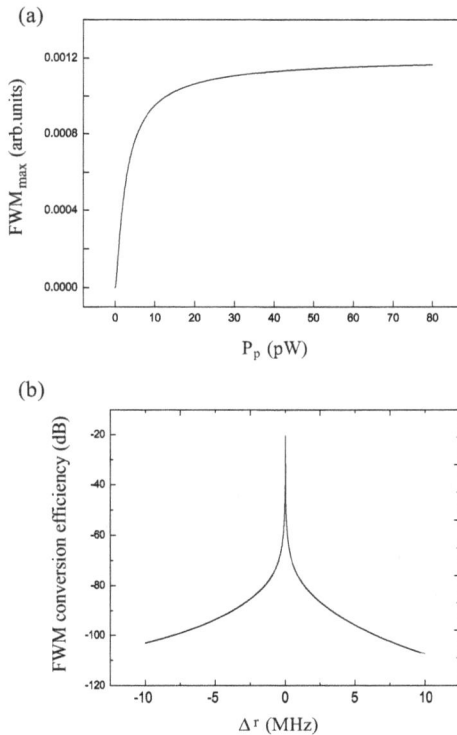

Fig. 9.12 (a) FWM intensity as a function of pump power P_p when $\Delta_p = \omega_n$ and $\Delta_r = 0$. (b) FWM conversion efficiency as a function of probe-cavity detuning when $\Delta_p = \omega_n$ and $P_p = 100$pW. Other parameters used are $\omega_c = 2\pi \times 7.5$GHz, $\omega_n = 2\pi \times 6.3$MHz, $\kappa = 2\pi \times 600$kHz, $g_0 = \pi \times 460$Hz, and $\gamma_n = 2\pi \times 30$Hz.

Figure 9.12(a) plots the peak value of FWM in the resonant region ($\Delta_r = 0$) as a function of the pump power with $\Delta_p = \omega_n$. With increasing of P_p, this value is greatly enhanced and finally reaches saturation at a low pump power, on the order of pW. Figure 9.12(b) shows the FWM conversion efficiency as a function of probe-cavity detuning, which demonstrates that the highest conversion efficiency is about -20dB when the pump power equals to 100pW. In this case, the conversion efficiency with lower pump power obtained here is higher than that was recently achieved in photonic crystal waveguides [Li (1996)]. This enhanced four-wave mixing based on OMIT will make a great contribution in the generation of non-classical states, such as squeezed states of microwave fields. The extremely low power with resonantly enhanced nonlinear coefficient may find applications in real communication networks. Recently, Reid *et al.* [Reid (1985)] and Slusher *et al.* [Slusher (1985)] have investigated the generation of squeezed states by FWM in an optical cavity in theory and experiment, respectively.

9.4   Mass sensing based on a circuit cavity electromechanical system

In previous chapters, we have discussed the mass sensing based on the coupled nanomechanical systems, due to their minuscule masses, high frequencies and high resonance quality factors. These mass sensors employ tracking the resonance frequency shifts of the resonators due to mass changes caused by accreted particles, in the presence of two optical fields. Because of the all-optical environment, temperature fluctuation noise and circuits noise have a promise to be avoided with pump-probe technique. Recently, it has been demonstrated experimentally that capacitive detection is less affected to noise than optical detection [Kim (2006)]. Capacitive detection can be realized by connecting NEMS resonators with standard microelectronics, such as electrical LC oscillator [Kim (2006)] and complementary metal-oxide-semiconductor (CMOS) circuitry [Forsen (2005)]. In this section, we propose a capacitive detection scheme for mass sensing based on a circuit cavity electromechanical system.

The optical (microwave) response of the cavity electromechanical system is modified in the presence of mechanical interactions, leading to effects such as parametric normal-mode splitting [Groblacher (2009)], optomechanically induced transparency [Agarwal (2010); Weis (2010); Teufel and Li (2011)] and slow light [Safavi-Naeini (2011)]. Because the thermal motion of the mechanical resonator creates an easily resolvable peak in the microwave noise spectrum, cavity electromechanical systems are quite suitable for mass sensing. In this section, the mass sensing scheme is based on the cavity electromechanical system where a flexural mode of a thin aluminum membrane is capacitively coupled to a superconducting microwave cavity which can be treated as a LC resonant circuit [Teufel and Donner (2011)]. When the cavity is driven by a strong microwave pump field and a weak probe field, the resonance frequency of the membrane can be easily determined from the probe transmission spectrum. Therefore, the accreted mass landing onto the membrane can be weighed conveniently according to the frequency shift.

The setup for mass sensing based on cavity electromechanical system is sketched in Fig.9.13(a), where the mechanical resonator is the top plate of a parallel-plate capacitor C and mass accretions upon it through the nozzle can be weighed. Specific experimental details of the cavity electromechanical system was demonstrated in Ref.[Teufel and Li (2011)]. Figure 9.13(b) is the diagram of measurement circuity where the superconducting microwave cavity is denoted by equivalent inductance L and equivalent capacitance C.

In previous chapters, we have discussed that the relationship between $\Delta\omega$ with the deposited mass Δm is given by

$$\Delta m = -\frac{2m_{\text{eff}}}{\omega_n}\Delta\omega = \Re^{-1}\Delta\omega, \tag{9.13}$$

where $\Re = (-2m_{\text{eff}}/\omega_n)^{-1}$ is defined as the mass responsivity. Here, we can determine the frequency shifts with high precision by the microwave spectroscopy measurement based on the circuit cavity electromechanical system.

We choose a realistic cavity electromechanical system to demonstrate the validity of our proposed mass sensing scheme. Parameters used in the numerical simulation are [Teufel and Li (2011)]: $\omega_c = 2\pi \times 7.5\text{GHz}$, $\omega_n = 2\pi \times 10.69\text{MHz}$, $\kappa = 2\pi \times 170\text{kHz}$, $\gamma_n = 2\pi \times 30\text{Hz}$, $Q_n = 360{,}000$, $g_0 = \pi \times 460\text{Hz}$, and the zero-point motion $x_{zp} = 4.1\text{fm}$. Therefore, we can obtain that the effective mass of the resonator is approximately equal to $m_{\text{eff}} = 46.8\text{pg}$ from the equation $x_{zp} = \sqrt{\hbar/2m_{\text{eff}}\omega_n}$.

According to Fig.9.4, we first measure the initial resonance frequency of mechanical resonator without landing any particles. As shown in Fig.9.14(a), we tune the frequency of the pump field to be resonant with the cavity resonance frequency ($\Delta_p = 0$) and scan the probe frequency across the cavity frequency, then the resonance frequency of the mechanical

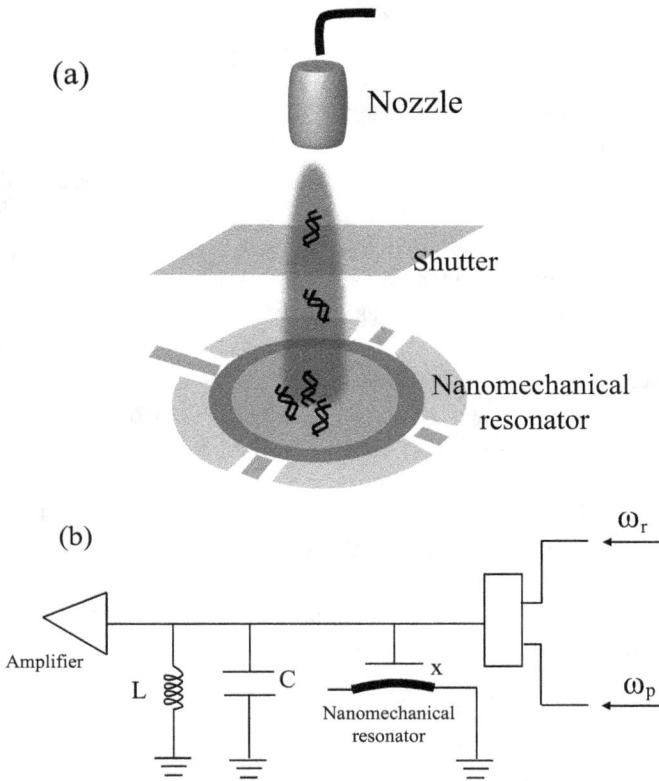

Fig. 9.13 (a) Schematic for mass sensing. A gas nozzle aperture provides a controlled flux of DNA molecules. The flux is gated by a mechanical shutter to provide calibrated, pulsed mass accretions upon the mechanical resonator. The resonator is the upper plate of a parallel-plate capacitor which is free to vibrated like a taut, circular drum. The experiments of our proposed mass sensing scheme should be done in situ within a cryogenically cooled, ultrahigh vacuum apparatus with base pressure below 10^{-10} Torr. (b) Diagram of the measurement circuity. The resonator is capacitively coupled to a superconducting microwave cavity denoted by equivalent inductance L and equivalent capacitance C. Two microwave tones ω_p and ω_r drive the coupled system simultaneously. The mechanical resonance is detected in a transmission scheme by using a low noise, cryogenic amplifier.

resonator can be observed in the probe transmission spectrum. In Fig.9.14(a), the resonance frequency of the mechanical resonator we selected is $\omega_n = 2\pi \times 10.69\text{MHz}$, then the steep peaks are located at $\Delta_r/2\pi = \pm 10.69\text{MHz}$. Figure 9.14(b) and Fig.9.14(c) shows the amplification of these two sidebands, where the linewidth equals to $\gamma_n/2\pi = 30\text{Hz}$.

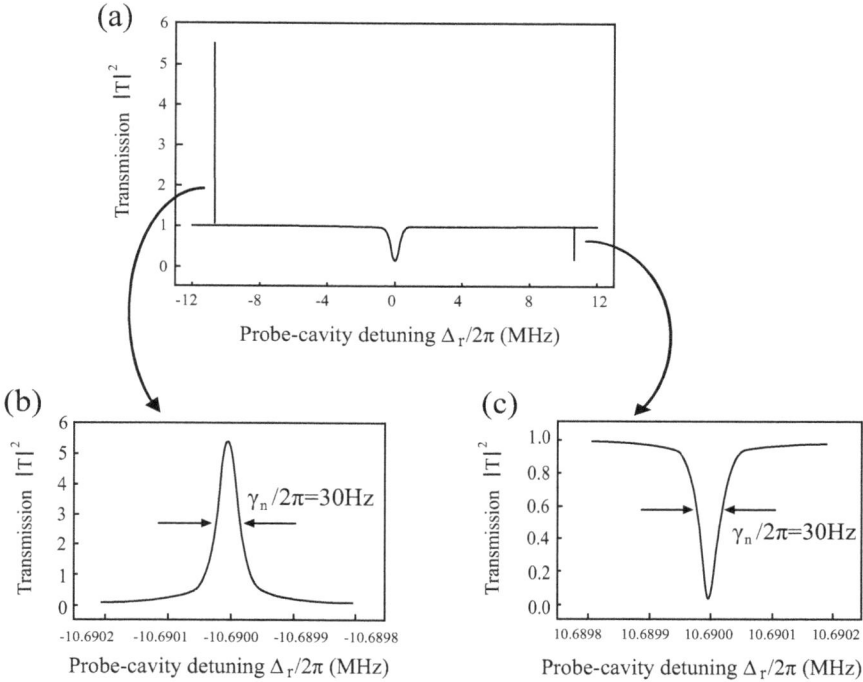

Fig. 9.14   (a) The magnitude of the probe transmission spectrum as a function of probe-cavity detuning Δ_r when $\Delta_p = 0$ and $P_p = 0.5$pW. (b) The amplification of the left sideband peak shown in (a). (c) The amplification of the right sideband peak shown in (a). Other parameters used are $\omega_c = 2\pi \times 7.5$GHz, $\omega_n = 2\pi \times 10.69$MHz, $\kappa = 2\pi \times 170$kHz, $\gamma_n = 2\pi \times 30$Hz, $Q_n = 360,000$, and $g_0 = \pi \times 460$Hz.

Next, we illustrate how to measure the mass of the particles landing on the resonator, according to Eq.(9.13). Here, we use the functionalized 1578 base pair long double-stranded deoxyribonucleic acid (dsDNA) molecules with mass $m_{\text{DNA}} \approx 1659$zg [Llic (2005)], and assume for simplicity that the mass adds uniformly to the mass of the overall resonator and changes the resonance frequency of the resonator by an amount. Figure 9.15 demonstrates the probe transmission spectrum as a function of Δ_r before and after a binding event of ~ 100 functionalized 1587bp DNA molecules in the vicinity of the mechanical resonance frequency. Resonance frequency shift $\Delta\omega = -2\pi \times 18.95$Hz is well resolved in the transmission spectrum after the adsorption of the DNA molecules due to the increased mass of the resonator. According to equation (9.13), the mass of the accreted DNA

(a)

Probe-cavity detuning $\Delta_r/2\pi$ (MHz)

(b)

Number of DNA molecules

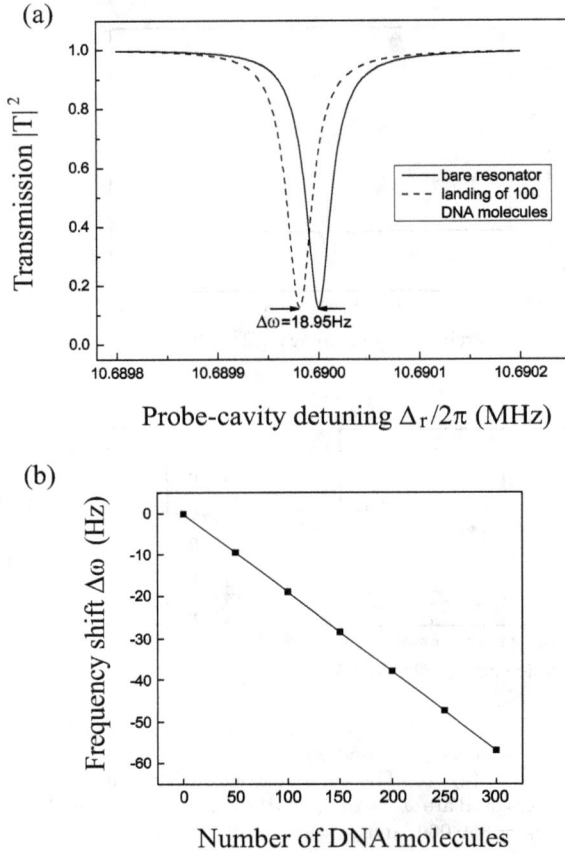

Fig. 9.15 (a) Plot of the probe transmission spectrum as a function of probe-cavity detuning before and after a binding event of ~ 100 functionalized 1578 base pair long double-stranded deoxyribonucleic acid (dsDNA) molecules. (b) The relationship between the frequency shifts and the number of DNA molecules. The parameters used are the same with Fig.9.14.

molecule can be: $\Delta m = -2m_{\text{eff}}\Delta\omega/\omega_n = 165900$zg, about the mass of 100 functionalized 1578 base pair long double-stranded deoxyribonucleic acid (dsDNA) molecules. Figure 9.15(b) demonstrates the direct linear relationship between the resonance frequency shifts and the number of DNA molecules landing on the resonator. The slope gives the mass responsivity of the resonator. Smaller mass and higher frequency of the resonator enable higher mass responsivity. For the cavity electromechanical system, the resonator we used for mass sensing is a nearly circular membrane of 100nm thick and 15μm in diameter, shown in Fig.9.13(a). The mass responsivity of the membrane for mass sensing is $|\Re| \approx 0.72$Hz/ag. Besides, mass resolution is given by the expression $\delta M \sim (m_{\text{eff}}/Q_n) \times 10^{-\text{DR}/20}$ (DR represents dynamic range) [Ekinci (2004)]. Here, the extremely high Q_n value combined with large dynamic range enable small mass resolution.

Bibliography

O. Arcizet, P.-F. Cohadon, T. Briant, M. Pinard, and A. Heidmann, Radiation-pressure cooling and optomechanical instability of a micromirror, Nature 444, 71 (2006).

I. Wilson-Rae, N. Nooshi, W. Zwerger, and T. J. Kippenberg, Theory of ground state cooling of a mechanical oscillator using dynamical backaction, Phys. Rev. Lett. 99, 093901 (2007).

F. Marquardt, Joe P. Chen, A. A. Clerk, and S. M. Girvin, Quantum theory of cavity-assisted sideband cooling of mechanical motion, Phys. Rev. Lett. 99, 093902 (2007).

C. Genes, D. Vitali, P. Tombesi, S. Gigan, and M. Aspelmeyer, Ground-state cooling of a micromechanical oscillator: Comparing cold damping and cavity-assisted cooling schemes, Phys. Rev. A 77, 033804 (2008).

T. J. Kippenberg and K. J. Vahala, Cavity optomechanics: Back-action at the mesoscale, 321, 1172 (2008).

S. Groblacher, K. Hammerer, M. R. Vanner, and M. Aspelmeyer, Observation of strong coupling between a micromechanical resonator and an optical cavity field, Nature 460, 724 (2009).

M. J. Woolley, A. C. Doherty, G. J. Milburn, and K. C. Schwab, Nanomechanical squeezing with detection via a microwave cavity, Phys. Rev. A 78, 062303 (2008).

C. A. Regal, J. D. Teufel, and K. W. Lehnert, Measuring nanomechanical motion with a microwave cavity interferometer, Nat. Phys. 4, 555 (2008).

J. D. Teufel, J. W. Harlow, C. A. Regal, and K. W. Lehnert, Dynamical back-action of microwave fields on a nanomechanical oscillator, Phys. Rev. Lett. 101, 197203 (2008).

T. Rocheleau, T. Ndukum, C. Macklin, J. B. Hertzberg, A. A. Clerk, and K. C. Schwab, Preparation and detection of a mechanical resonator near the ground state of motion, Nature 463, 72 (2010).

A. D. Armour, M. P. Blencowe, and K. C. Schwab, Entanglement and decoherence of a micromechanical resonator via coupling to a Cooper-pair box, Phys. Rev. Lett. 88, 148301 (2002).

A. Wallraff, D. I. Schuster, A. Blais, L. Frunzio, R.-S. Huang, J. Majer, S. Kumar, S. M. Girvin, and R. J. Schoelkopf, Strong coupling of a single photon to a superconducting qubit using circuit quantum electrodynamics, Nature (London) 431, 162 (2004).

D. Vitali, P. Tombesi, M. J. Woolley, A. C. Doherty, and G. J. Milburn, Entangling a nanomechanical resonator and a superconducting microwave cavity, Phys. Rev. A 76, 042336 (2007).

L. Tian, M. S. Allman, R. W. Simmonds, Parametric coupling between macroscopic quantum resonators, New J. Phys. 10, 115001 (2008).

J. B. Hertzberg, T. Rocheleau, T. Ndukum, M. Savva, A. A. Clerk, and K. C. Schwab, Back-action-evading measurements of nanomechanical motion, Nat. Phys. 6, 213 (2010).

F. Xue, Y. D. Wang, Y. X. Liu, and F. Nori, Cooling a micromechanical beam by coupling it to a transmission line, Phys. Rev. B 76, 205302 (2007).

Y. Li, Y. D. Wang, F. Xue, and C. Bruder, Quantum theory of transmission line resonator-assisted cooling of a micromechanical resonator, Phys. Rev. B 78, 134301 (2008).

J. D. Teufel, C. A. Regal, and K. W. Lehnert, Prospects for cooling nanomechanical motion by coupling to a superconducting microwave resonator, New J. Phys. 10, 095002 (2008).

A. Schliesser, Cavity optomechanics and optical frequency comb generation with silica whispering-gallery-mode microresonators, Thesis, Ludwig-Maximilians-Universität München (2009); http://edoc.ub.uni-muenchen.de/10940.

C. Jiang, B. Chen, and K. D. Zhu, Controllable nonlinear responses in a cavity electromechanical system, J. Opt. Soc. Am. B 29, 220 (2012).

C. Jiang, B. Chen, J. J. Li and K. D. Zhu, Mass sensing based on a circuit cavity electromechanical system, J. Appl. Phys. 110, 083107 (2011).

S. Weis, R. Rivière, S. Deléglise, E. Gavartin, O. Arcizet, A. Schliesser, and T. J. Kippenberg, Optomechanically induced transparency, Science 330, 1520 (2010).

L. V. Hau, S. E. Harris, Z. Dutton, C. H. Behroozi, Light speed reduction to 17 metres per second in an ultracold atomic gas, Nature 397, 594 (1999).

B. I. Greence, J. F. Mueller, J. Orenstein, D. H. Rapkine, S. S. Rink, and M. Thakur, Phonon-mediated optical nonlinearity in polydiacetylene, Phys. Rev. Lett. 61, 325 (1988).

R. W. Boyd, Nonlinear optics, (Academic Press, Amsterdam), pp. 297 (2008).

C. W. Gardiner and P. Zoller, Quantum noise, (Springer, NY) (2004).

S. Huang and G. S. Agarwal, Normal-mode splitting and antibunching in Stokes and anti-Stokes processes in cavity optomechanics: Radiation-pressure-induced four-wave-mixing cavity optomechanics, Phys. Rev. A 81, 033830 (2010).

J. Li, L. O'Faolain, I. H. Rey, and T. F. Krauss, Four-wave mixing in photonic crystal waveguides: slow light enhancement and limitations, Opt. Express 19, 4458 (2011).

S. Gupta, K. L. Moore, K. W. Murch, and D. M. Stamper-Kurn, Cavity nonlinear optics at low photon numbers from collective atomic motion, Phys. Rev. Lett. 99, 213601 (2007).

F. Brennecke, S. Ritter, T. Donner, and T. Esslinger, Cavity optomechanics with a Bose-Einstein condensate, Science, 322, 235 (2008).

R. Kanamoto and P. Meystre, Optomechanics of a quantum-degenerate Fermi gas, Phys. Rev. Lett. 104, 063601 (2010).

B. P. Venkatesh, J. Larson, and D. H. J. ODell, Band-structure loops and multi-stability in cavity QED, Phys. Rev. A 83, 063606 (2011).

R. Ghobadi, A. R. Bahrampour, and C. Simon, Quantum optomechanics in the bistable regime, Phys. Rev. A 84, 033846 (2011).

A. Lezama, S. Barreiro, and A. M. Akulshin, Electromagnetically induced absorption, Phys. Rev. A 59, 4732 (1999).

B. R. Mollow, R. J. Glauber, Quantum theory of prametric amplification, I Phys. Rev. 160, 1076 (1967).

P. Verlot, A. Tavernarakis, T. Briant, P.-F. Cohadon, and A. Heidmann, Back-action amplification and quantum limits in optomechanical measurements, Phys. Rev. Lett. 104, 133602 (2010).

V. B. Braginsky, S. E. Strigin, and S. P. Vyatchanin, Parametric oscillatory instability in Fabry-Perot interferometer, Phys. Lett. A 287, 331 (2001).

H. J. Kimble, The quantum internet, Nature 453, 1023 (2008).

T. Aoki, A. S. Parkins, D. J. Alton, C. A. Regal, Barak Dayan, E. Ostby, K. J. Vahala, and H. J. Kimble, Efficient routing of single photons by one atom and a microtoroidal cavity, Phys. Rev. Lett. 102, 083601 (2009).

M. Fleischhauer, A. Imamoğlu, and J. P. Marangos, Electromagnetically induced transparency: Optics in coherent media, Rev. Mod. Phys. 77, 633 (2005).

M. Bajcsy, S. Hofferberth, V. Balic, T. Peyronel, M. Hafezi, A. S. Zibrov, V. Vuletic, and M. D. Lukin, Efficient all-optical switching using slow light within a hollow fiber, Phys. Rev. Lett. 102, 203902 (2009).

M. A. Hall, J. B. Altepeter, and P. Kumar, Ultrafast switching of photonic entanglement, Phys. Rev. Lett. 106, 053901 (2011).

G. S. Agarwal and S. M. Huang, Optomechanical systems as single-photon routers, Phys. Rev. A 85, 021801 (2012).

O. Astafiev, A. M. Zagoskin, A. A. Abdumalikov, Y. A. Pashkin, T. Yamamoto, K. Inomata, Y. Nakamura, and J. S. Tsai, Resonance fluorescence of a single artificial atom, Science 327, 840 (2010).

A. A. Abdumalikov, O. Astafiev, A. M. Zagoskin, Y. A. Pashkin, Y. Nakamura, and J. S. Tsai, Electromagnetically induced transparency on a single artificial atom, Phys. Rev. Lett. 104, 193601 (2010).

I. Hoi, C. M. Wilson, G. Johansson, T. Palomaki, B. Peropadre, and P. Dsing, Demonstration of a single-photon router in the microwave regime, Phys. Rev. Lett. 107, 073601 (2011).

J. D. Teufel, T. Donner, Dale Li, J. W. Harlow, M. S. Allman, K. Cicak, A. J. Sirois, *et al.*, Sideband cooling of micromechanical motion to the quantum ground state, Nature 475, 359 (2011).

S. Diez, C. Schmidt, R. Ludwig, H. G. Weber, K. Obermann, S. Kindt, I. Koltchanov, K. Petermann, Four-wave mixing in semiconductor optical amplifiers for frequency conversion and fast optical switching, IEEE J. Sel. Top. Quant. 3, 1131 (1997).

K. Kitayama, Highly stabilized millimeter-wave generation by using fiber-optic frequency-tunable comb generator, J. Lightwave Technol. 15, 883 (1997).

A. Wiberg, P. P. Millán, M. V. Andrés, and P. O. Hedekvist, Microwave-photonic frequency multiplication utilizing optical four-wave mixing and fiber bragg gratings, J. Lightwave Technol. 24, 329 (2006).

S. E. Harris, J. E. Field, and A. Imamoğlu, Nonlinear optical processes using electromagnetically induced transparency, Phys. Rev. Lett. 64, 1107 (1990).

K.-J. Boller, A. Imamoğlu, and S. E. Harris, Observation of electromagnetically induced transparency, Phys. Rev. Lett. 66, 2593 (1991).

Y. Li and M. Xiao, Enhancement of nondegenerate four-wave mixing based on electromagnetically induced transparency in rubidium atoms, Opt. Lett. 21, 1064 (1996).

H. Schmidt and A. Imamoğlu, Giant Kerr nonlinearities obtained by electromagnetically induced transparency, Opt. Lett. 21, 1936 (1996).

B. S. Ham, M. S. Shahriar, and P. R. Hemmer, Enhanced nondegenerate four-wave mixing owing to electromagnetically induced transparency in a spectral hole-burning crystal, Opt. Lett. 22, 1138 (1997).

J. D. Teufel, D. Li, M. S. Allman, K. Cicak, A. J. Sirois, J. D. Whittaker, and R. W. Simmonds, Circuit cavity electromechanics in the strong-coupling regime, Nature 471, 204 (2011).

G. S. Agarwal and S. M. Huang, Electromagnetically induced transparency in mechanical effects of light, Phys. Rev. A 81, 041803(R) (2010).

A. H. Safavi-Naeini, T. P. Mayer Alegre, J. Chan, M. Eichenfield, M. Winger, Q. Lin, J. T. Hill, D. E. Chang, and O. Painter, Electromagnetically induced transparency and slow light with optomechanics, Nature 472, 69 (2011).

M. D. Reid and D. F. Walls, Generation of squeezed states via degenerate four-wave mixing, Phys. Rev. A 31, 1622 (1985).

R. E. Slusher, L. W. Hollberg, B. Yurke, J. C. Mertz, and J. F. Valley, Observation of squeezed states generated by four-wave mixing in an optical cavity, Phys. Rev. Lett. 55, 2409 (1985).

S. J. Kim, T. Ono, and M. Esashi, Capacitive resonant mass sensor with frequency demodulation detection based on resonant circuit, Appl. Phys. Lett. 88, 053116 (2006).

E. Forsen, G. Abadal, S. G. Nilsson, J. Teva, J. Verd, R. Sandberg, W. Svendsen, F. P. Murano, J. Esteve, E. Figueras, F. Campabadal, L. Montelius, N. Barniol, and A. Boisen, Ultrasensitive mass sensor fully integrated with complementary metal-oxide-semiconductor circuitry, Appl. Phys. Lett. 87, 043507 (2005).

A. H. Safavi-Naeini, T. P. Mayer Alegre, J. Chan, M. Eichenfield, M. Winger, Q. Lin, J. T. Hill, D. E. Chang, and O. Painter, Electromagnetically induced transparency and slow light with optomechanics, Nature 472, 69 (2011).

B. Llic, Y. Yang, K. Aubin, R. Reichenbach, S. Krylo, and H. G. Craighead, Enumeration of DNA molecules bound to a nanomechanical oscillator, Nano Lett. 5, 925 (2005).

K. L. Ekinci, Y. T. Tang, and M. L. Roukes, Ultimate limits to inertial mass sensing based upon nanoelectromechanical systems, J. Appl. Phys. 95, 2682 (2004).

Chapter 10

A Hybrid Optomechanical System Based on Quantum Dot and DNA Molecules

Optical detection of DNA molecules provides a platform for automated biological experiments and ultrasensitive cell detections, i.e., optical imaging, optical sensing, immunoassay, etc. [Liedl (2012); Weizmann (2006)]. These researches focus on the inorganic/organic hybrid DNA systems in biological labels, cell tracking, and monitoring response to therapeutic agents [Wu (2003); Baur (2010)]. Quantum dot is one of the best candidates for the hybrid biological system, due to its unique size-dependent, narrow, symmetric, bright, and stable fluorescence [Fu (2005)]. Furthermore, conjugating to other bio-molecules can not change the optical behaviors of the quantum dot, which will offer a platform for the photon-limited studies and tracking experiments in the future.

Researchers have shown that QD-linked conjugates are promising nanoprobes for studying biological activities and that they will have a wide rage of applications in biological staining and medical diagnosis [Hu (2010); Zou (2010); Medintz (2005, 2003)]. Consequently, some systems used quantum dots linked with DNA probe to explore biomolecule targets [Gilroy (2010); Kim (2009); Michalet (2005)]. Recently, Kim and coauthors [Kim (2006)] have presented a sensitive DNA detection protocol using quantum dots (QDs) and magnetic beads (MBs) for large volume samples. In their study, the fluorescent intensity was proportional to concentration of the target DNAs, and the low copies of target DNAs was successfully detected. To make highly sensitive DNA-based sensors, Zhou et al. [Zhou (2008)] have reported the preparation of a compact, functional quantum dot-DNA conjugate, which is capable of specific detection of nanomolar unlabeled complimentary DNA at low DNA probe/QD copy numbers via a "signal-on" fluorescence resonance energy transfer (FRET) process. Therefore, the study of the spectroscopy and the dynamic behavior of DNA molecules is important for understanding the mechanisms of DNA biological function and open new possibilities in biomedicine technique such as fluorescent probe science and analytical chemistry [Zhang (2005); Artemyev (2009)].

In this chapter, we propose a generalized optomechanical system based on DNA enhanced signal spectroscopy in a quantum dot (QD) coupled by DNA molecules system. Quantum dots are potential candidate for high resolution imaging techniques, such as electron microscopy and x-ray microscopy [Fu (2005)]. However, because of the autofluorescence of background, the coated chemicals and the copy number of the target, the fluorescence emission efficiencies remain big challenge in QDs-linked biomedicine sensors [Zhou (2008)]. Therefore, we focus on the enhanced signal spectroscopy of quantum dot by coupled DNA molecules using two-optical technique. When applying a strong control laser, the coupled QD-DNA system will become transparent, and one can scan the second signal field across the exciton frequency of the QD, then the output signal laser will be amplified without any additional distortion. This amplification results in the enhanced luminescence of QD in optical domain. With better control of the emission efficiency of QD by tuning the intensity of control laser, this two-optical technique will hopefully be possible to distinguish tumor cells from normal ones. Furthermore, the vibrational frequency of DNA molecule and the coupling strength between QD and DNA can be measured simultaneously.

10.1   Model and theory

A biological semiconductor neutral quantum dot is coupled to some DNA molecules, which is shown in Fig.10.1(a). We apply two optical fields on this coupled system, one is strong control field, and the other is weak signal field. The energy levels of the semiconductor neutral quantum dot can be treated as two-level system, which consist of the ground state $|g\rangle$ and the first excited state (single exciton) $|ex\rangle$ (Fig.10.1(b). This two-level exciton can be characterized by the pseudospin -1/2, operators $S\pm$ and S_z. Furthermore, Xu *et al.* [Xu (2007, 2008)] has already detected the coherent optical spectroscopy of a strongly driven quantum dot. As usual, the Hamiltonian of the exciton in a single QD can be described as

$$H_{QD} = \hbar\omega_{\text{ex}}S^z, \tag{10.1}$$

where ω_{ex} is the exciton frequency of the quantum dot.

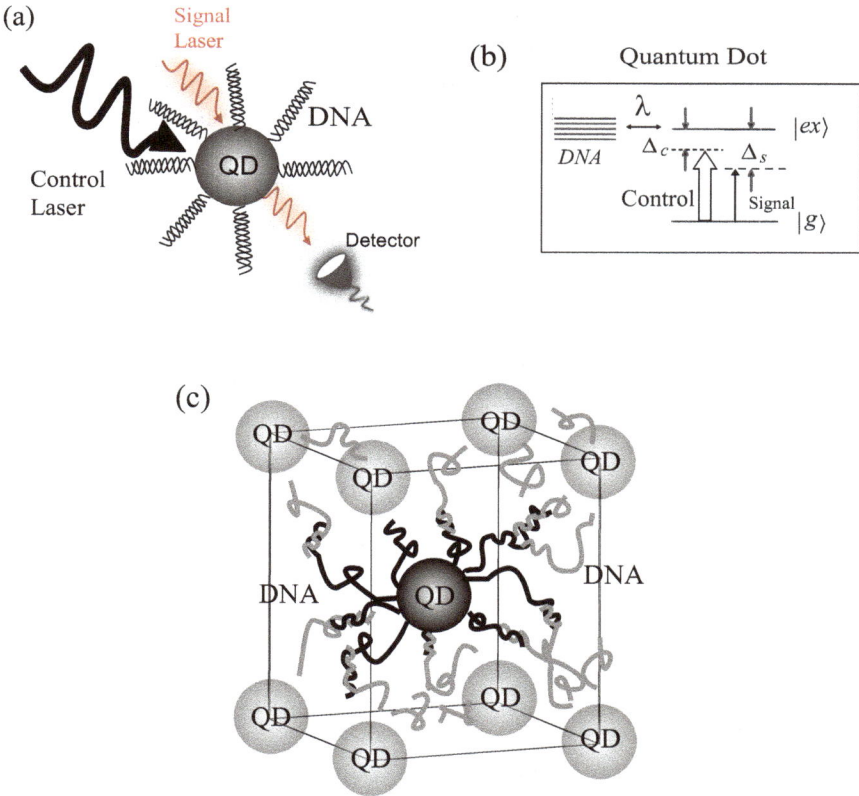

Fig. 10.1   (a) Schematic of a quantum dot coupled to DNA molecules in the simultaneous presence of two optical fields. (b) The energy levels of quantum dot when dressing the vibrational modes of DNA molecules. (c) The model of our QD-DNA proposal, where many DNA molecules coupled to one quantum dot.

We treat the DNA molecules as harmonic oscillators. Therefore, the Hamiltonian of the DNA molecules can be described by the position operator q and momentum operator p. They obey the commutation relation $[q, p] = i\hbar$ [Van Zandt (1986)], and then

$$H_D = \sum_{i=1}^{n} \left(\frac{p_i^2}{2m_i} + \frac{1}{2} m_i \omega_i^2 q_i^2 \right),$$ \hfill (10.2)

where m_i and ω_i are the mass and vibrational frequency of DNA molecule, respectively.

Next, we consider the vibrational modes of DNA molecules when coupling with a single quantum dot. Here, the vibrational modes of DNA
molecules can be treated as harmonic vibration of spring oscillators. In
a large volume of aqueous solution (or the unlimited solution), all the vibrational modes of DNA molecules are strongly attenuated. Otherwise, in
a small volume of aqueous solution, the longitudinal vibrational modes of
DNA molecules decayed slowly, while the other modes remain strong attenuation [Dorfman (1984)]. In this case, for the small volume solution, only
the longitudinal modes of DNA molecules can be considered. As shown in
Fig.10.1(c), the flexion of DNA molecules (known as longitudinal strain)
produces extensions and compression, which can modify the energy of the
electronic states of QD via deformation potential coupling [Edwards (1984,
1985)]. Then the Hamiltonian of the vibrational modes of DNA molecules
coupled to the QD can be described by

$$H_{QD-D} = \hbar S^z \sum_{i=1}^{n} M_i q_i, \qquad (10.3)$$

where M_i is the coupling strength between the QD and the *ith* DNA.
It is should be noted that due to the dilute aqueous solution of DNA
molecules, here we do not consider the effect of the coupling between the
DNA molecules although it may be significant in the dense aqueous solutions [Dorfman (1984)].

The Hamiltonian of the quantum dot coupled to the strong control field
and weak signal field is described as

$$\begin{aligned}
H_{QD-f} = &-\mu(E_c S^+ e^{-i\omega_c t} + E_c^* S^- e^{i\omega_c t}) \\
&-\mu(E_s S^+ e^{-i\omega_s t} + E_s^* S^- e^{i\omega_s t}),
\end{aligned} \qquad (10.4)$$

where μ is the electric dipole moment of the exciton, ω_c (ω_s) is the frequency
of the control field (signal field), and E_c (E_s) is the slowly varying envelope
of the control field (signal field). Therefore, we obtain the total Hamiltonian
of the coupled QD-DNA in the presence of two optical fields as follows

$$\begin{aligned}
H = &\, H_{QD} + H_D + H_{QD-D} + H_{QD-f} \\
= &\, \hbar\omega_{\text{ex}} S^z + \sum_{i=1}^{n}\left(\frac{p_i^2}{2m_i} + \frac{1}{2}m_i\omega_i^2 q_i^2\right) + \hbar S^z \sum_{i=1}^{n} M_i q_i \\
&- \mu(E_c S^+ e^{-i\omega_c t} + E_c^* S^- e^{i\omega_c t}) - \mu(E_s S^+ e^{-i\omega_s t} + E_s^* S^- e^{i\omega_s t}).
\end{aligned} \qquad (10.5)$$

In a rotating frame at the control field ω_c, the total Hamiltonian of the system reads as

$$
H = \hbar\Delta_c S^z + \sum_{i=1}^{n}\left(\frac{p_i^2}{2m_i} + \frac{1}{2}m_i\omega_i^2 q_i^2\right) + \hbar S^z Q
$$
$$
-\hbar(\Omega_c S^+ + \Omega_c^* S^-) - \mu(E_s S^+ e^{-i\delta t} + E_s^* S^- e^{i\delta t}), \qquad (10.6)
$$

where $\Delta_c = \omega_{\text{ex}} - \omega_c$, $Q = \sum_{i=1}^{n} M_i q_i$, Ω_c is the Rabi frequency of the control field, and $\delta = \omega_s - \omega_c$ is the detuning between the signal field and the control field.

According to the Heisenberg equation of motion, we can obtain the following equations:

$$
\frac{d}{dt}S^z = -\Gamma_1\left(S^z + \frac{1}{2}\right) + i\Omega_c(S^+ - S^-) + \frac{i\mu E_s e^{-i\delta t}}{\hbar}S^+ - \frac{i\mu E_s^* e^{i\delta t}}{\hbar}S^-,
$$
$$
(10.7)
$$

$$
\frac{d}{dt}S^- = -[i(\Delta_c + Q) + \Gamma_2]S^- - 2i\Omega_c S^z - \frac{2i\mu E_s e^{-i\delta t}}{\hbar}S^z, \qquad (10.8)
$$

$$
\frac{d^2}{dt^2}Q + \frac{1}{\tau_D}\frac{d}{dt}Q + \omega_D^2 Q = -\lambda\omega_D^2 S^z, \qquad (10.9)
$$

where $\lambda = \sum_{i=1}^{n}\hbar M_i^2/(m_i\omega_D^2)$ corresponds to the QD-DNA coupling strength, ω_D is the DNA longitudinal vibrational modes, Γ_1 and Γ_2 are the exciton relaxation rate and dephasing rate, respectively, τ_D is the vibrational lifetime of DNA molecule [Donega (2006)]. In order to solve these equations, we first take the semiclassical approach by factorizing the DNA molecule and exciton degrees of freedom, i.e., $\langle QS^z\rangle = \langle Q\rangle\langle S^z\rangle$, in which any entanglement between these systems should be ignored. And then we make the following ansatz [Boyd (2008)]

$$
S^-(t) = S_0 + S_+ e^{-i\delta t} + S_- e^{i\delta t}, \qquad (10.10)
$$
$$
S^z(t) = S_0^z + S_+^z e^{-i\delta t} + S_-^z e^{i\delta t}, \qquad (10.11)
$$
$$
Q(t) = Q_0 + Q_+ e^{-i\delta t} + Q_- e^{i\delta t}. \qquad (10.12)
$$

Upon substituting these equations to Eqs.(10.7)-(10.9) and working to the lowest order in E_s, but to all orders in E_c, we finally obtain the linear optical susceptibility S_+ in the steady state as the following solution

$$
\chi(\omega_s)_{\text{eff}}^{(1)} = \frac{\mu S_+}{E_s} = \frac{\mu^2}{\hbar\Gamma_2}\chi(\omega_s), \qquad (10.13)
$$

where the dimensionless susceptibility $\chi(\omega_s)$ is given by

$$\chi(\omega_s) = \frac{w_0}{f(\delta_0)} \times \{e_1 e_2[(2i + \delta_0)(e_1 + \delta_0) - 2\Omega_{co}^2]$$
$$- e_2 \Omega_{co}^2 \lambda_0 \eta w_0 + \Omega_{co}^2 (2e_2 - \lambda_0 \eta w_0)(e_1 + \delta_0)\}, \quad (10.14)$$

where the function $f(\delta_0)$ and auxiliary function $\eta(\omega_s)$ are given by

$$f(\delta_0) = (e_2 + \delta_0)\{e_1 e_2[(2i + \delta_0)(e_1 + \delta_0) - 2\Omega_{co}^2]$$
$$- e_2 \Omega_{co}^2 \lambda_0 \eta w_0\} - e_1 \Omega_{co}^2 (2e_2 - \lambda_0 \eta w_0)(e_1 + \delta_0), \quad (10.15)$$

$$\eta(\omega_s) = \frac{\omega_{D0}^2}{\omega_{D0}^2 - \delta_0^2 - i\delta_0/\tau_{D0}}, \quad (10.16)$$

where

$$e_1 = i + \Delta_{c0} - \lambda_0/(2w_0), \quad (10.17)$$
$$e_2 = i - \Delta_{c0} + \lambda_0/(2w_0), \quad (10.18)$$
$$\delta_0 = \delta/\Gamma_2, \quad (10.19)$$
$$\Omega_{c0} = \Omega_c/\Gamma_2, \quad (10.20)$$
$$\lambda_0 = \lambda/\Gamma_2, \quad (10.21)$$
$$\omega_{D0} = \omega_D/\Gamma_2, \quad (10.22)$$
$$\tau_{D0} = \tau_D \Gamma_2, \quad (10.23)$$
$$\Delta_{c0} = \Delta_c/\Gamma_2, \quad (10.24)$$
$$w_0 = 2S_0^z, \quad (10.25)$$
$$\Gamma_1 = 2\Gamma_2. \quad (10.26)$$

The population inversion (w_0) of the exciton is determined by the following equation

$$(w_0 + 1)[(\Delta_{c0} - \frac{\lambda_0 w_0}{2})^2 + 1] + 2\Omega_{c0}^2 w_0 = 0. \quad (10.27)$$

10.2   Coherent optical spectrum

Equation (10.27) has three roots, which demonstrate the characteristic of the optical multistability in a coupled quantum dot and DNA molecules [Larson (2008); Gupta (2007)]. Figure 10.2 shows the steady-state value of the population inversion w_0 as a function of (a) control-exciton detuning, and (b) control Rabi frequency. Here, the bistable steady state proves that the coupled quantum dot and DNA molecules can act as a generalized

(a)

(b)

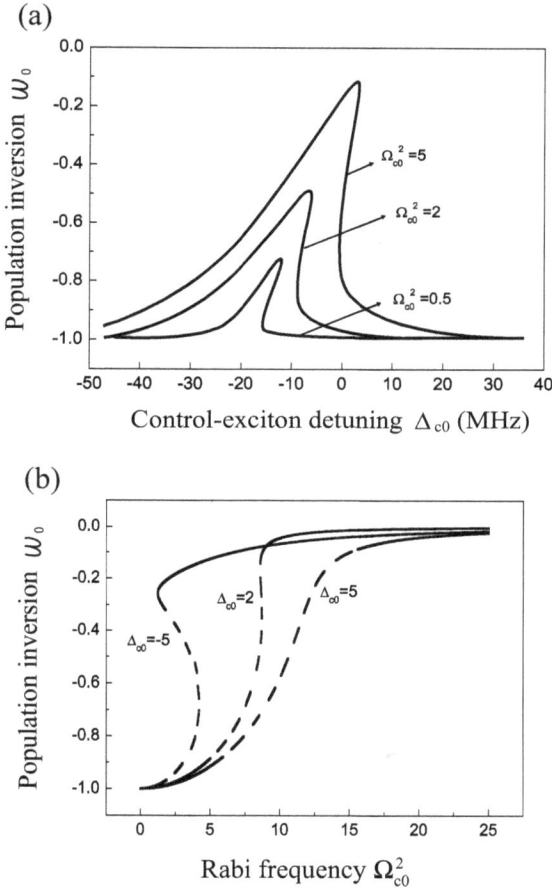

Fig. 10.2   Population inversion w_0 as a function of (a) control-exciton detuning and (b) Rabi frequency of the control laser. The dashed curves and solid curves correspond to the unstable states and stable states, respectively, where $\lambda_0 = 40$.

optomechanical system. The exciton-DNA molecule interaction is in analogy with optical cavity coupling with mechanical resonator in cavity optomechanical system. Figure 10.2(a) displays that the bistable is more obvious when increasing the control power gradually. And the bistable behavior is enhanced at the negative detuning $-\Delta_{co}$, as shown in Fig.10.2(b).

In the presence of a strong control field and a weak signal field, the coupled QD and DNA molecules system exhibits specific optical features. As shown in Fig.10.1(c), one QD is linked with many distorted DNA molecules.

However, these distorted DNA molecules can be extended into linear form while applying fluid force or electromagnetic field [Marko (1995)]. Here, we should notice that the length of DNA molecules is the key factor for the longitudinal vibrational frequency of the DNA molecules. During the theoretical simulation, we select the vibrational frequency and the lifetime of DNA molecule are $\omega_D = 32\text{GHz}$ and $\tau_D = 3\text{ns}$, respectively [Edwards (1985); Yuan (2008); Gill (2005); Adair (2002)]. The decay time of the quantum dot is 6fs [Amdursky (2010)], which corresponds to $\Gamma_1 = 160\text{THz}$.

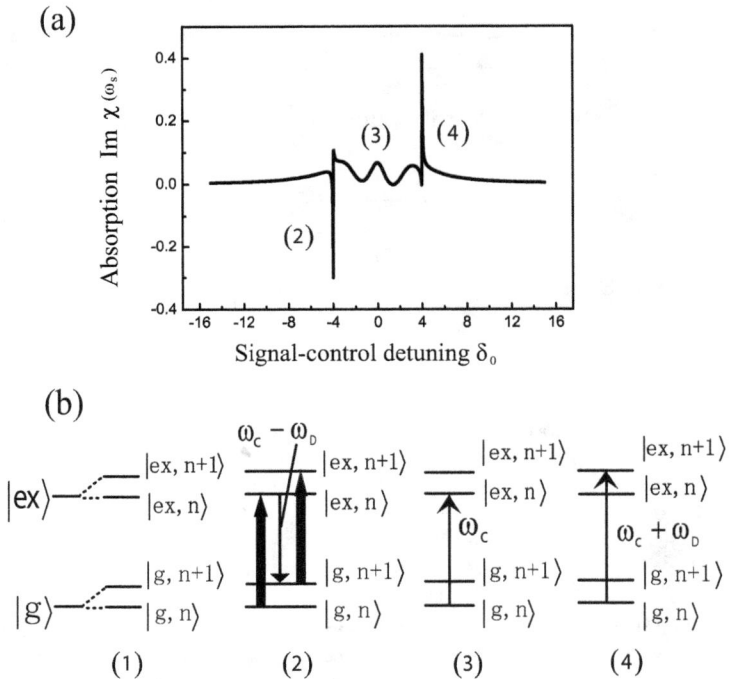

Fig. 10.3 The signal absorption spectrum in the presence of a strong control field. The parameters used are $\Omega_{c0}^2 = 2$, $\Delta_{c0} = 0$, $\lambda_0 = 0.5$, $\omega_{D0} = 4$, $\tau_{D0} = 33$. (b) The energy levels of exciton when coupling with the vibration of DNA molecules. Part (2), (3), and (4) correspond to the transitions of the left peak, center feature and right peak, respectively, shown in (a). $|n\rangle$ denotes the number states of the DNA vibrations.

The signal absorption spectrum as a function of control-signal detuning is shown in Fig.10.3(a), where a small peak is located at the center and two steep peaks are located at the both sides. The energy levels and the transitions corresponding to these three feature are shown in Fig.10.3(b), where part (1) shows the dressed states of exciton when QD interaction with DNA molecules; part (2) corresponds the transitions of the left negative peak shown in (a); part (3) and part (4) describe the transitions of the middle peak and the right absorbed peak shown in (a), respectively. Part (2) is the origin of DNA vibrational mode induced three-photon resonance, where the electron makes a transition from the lowest dressed level $|g, n\rangle$ to the highest dressed level $|\text{ex}, n+1\rangle$ by the simultaneous absorption of two control photons and emission of a photon at $\omega_c - \omega_D$. This magnified process can amplify a wave at $\delta = -\omega_D$. Part (3) is the DNA stimulated Rayleigh resonance, which corresponds to a transition from the lowest dressed level $|g, n\rangle$ to the dressed level $|\text{ex}, n\rangle$. Part (4) is the usual absorption resonance as modified by the ac-Stark effect, which is an optical absorption process.

10.3 Vibrational frequency measurement of DNA molecule

Figure 10.4 plots the signal absorption spectrum as a function of detuning Δ_s, with different resonance frequency of DNA molecules. Clearly, the two steep peaks at the both sides correspond to the resonance frequency of DNA molecule, i.e., when the frequency of DNA is 32GHz, the two steep peaks

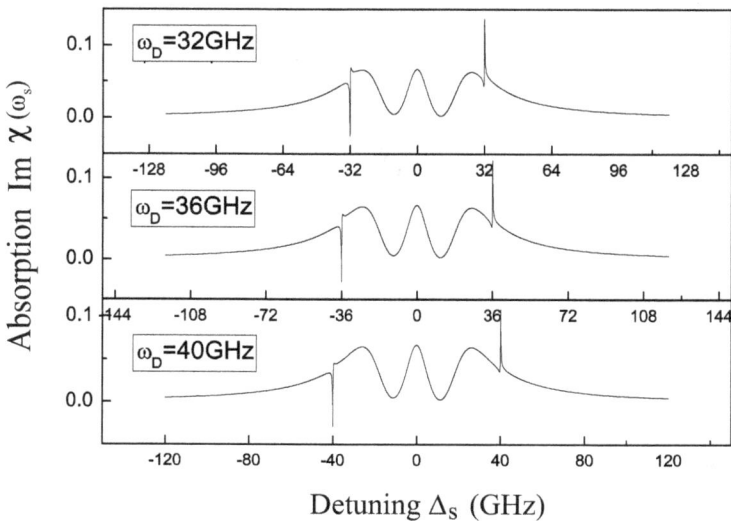

Fig. 10.4 The absorption spectrum of signal laser with different frequencies of DNA molecule. The parameters used are $I_c = 1.7\text{W}/\text{cm}^2$, $\Delta_c = 0$, $\lambda = 0.8\text{GHz}$, $\tau_D = 3\text{ns}$.

will be located at $\Delta_s = \pm 32$GHz (the top curve of Fig.10.4). The middle curve and bottom curve of Fig.10.4 shows that the location of two peaks can be changed with different frequencies of DNA molecules. In this case, this control-signal technique offers a simple and effective method for the detection of DNA resonance frequency. That it, by fixing the control field detuning $\Delta_c = 0$ and scanning the second signal field across the resonance frequency of exciton, the output signal field while passing through the coupled QD-DNA will exhibits the optical characteristics like Fig.10.4. The location of two steep peaks at the both sides of signal absorption spectrum just corresponds to the resonance frequency of DNA molecule.

10.4   Coupling strength determination between quantum dot and DNA molecule

If tuning the control detuning Δ_c on the frequency of DNA resonance, i.e., $\Delta_c = \omega_D$, the coupling strength between quantum dot and DNA molecule can be measured using the two-optical technique. The curves in Fig.10.5(a) showing the splitting peaks exhibit that different coupling strengths result in different peak splitting around $\Delta_s = 0$. This is due to DNA vibration induced coherent population oscillation which makes a deep hole at $\Delta_s = 0$ in the signal absorption spectrum as $\delta = \omega_D$. However, in the absence of the coupling strength between the DNA molecules and the quantum dot, the splitting peaks become a absorption curve, which is shown in the solid curve in Fig.10.5(a). We plot the relationship between the coupling strength of QD-DNA molecules and the peak splitting distance as shown in Fig.10.5(b), which offers a platform to measure the coupling strength between QD and DNA molecules. That is, after detection of the peak splitting distance in the signal absorption spectrum, one can known the coupling strength of QD-DNA, according to the linear relationship displayed in Fig.10.5(b).

10.5   A protocol of tumor discrimination

The left negative peak shown in Fig.10.4 means the output signal light can be amplified at $\Delta_s = -\omega_D$ and $\Delta_c = 0$. This also provides a potential application to discriminate the biological molecules, using two optical fields. Figure 10.6(a) shows the signal transmission spectrum with different control laser intensities, which indicates that the amplification of the signal laser is enhanced with the increasing of control laser intensity. Figure 10.6(b) shows the relationship between the signal transmission and the intensity of control field, which means the amplification of the output signal field can be modulated by the input control field.

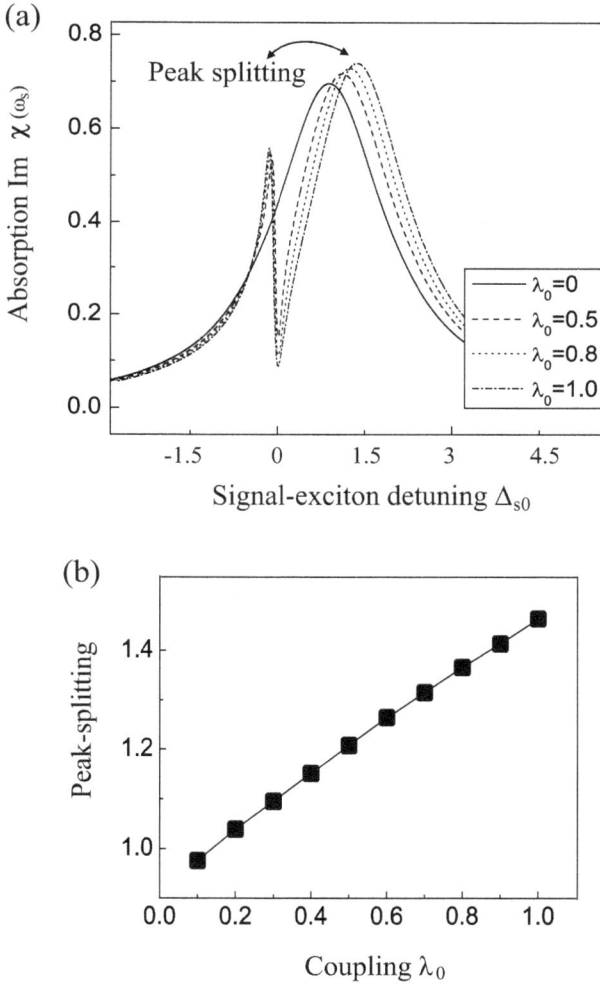

Fig. 10.5 The absorption spectrum of signal laser as a function of detuning between signal laser and exciton for four difference coupling of QD-DNA. (b) The linear relationship between the peak splitting and the coupling strength of QD-DNA. The parameters used are $\Omega_{c0}^2 = 2$, $\Delta_{c0} = 4$, $\omega_{D0} = 4$, $\tau_{D0} = 33$.

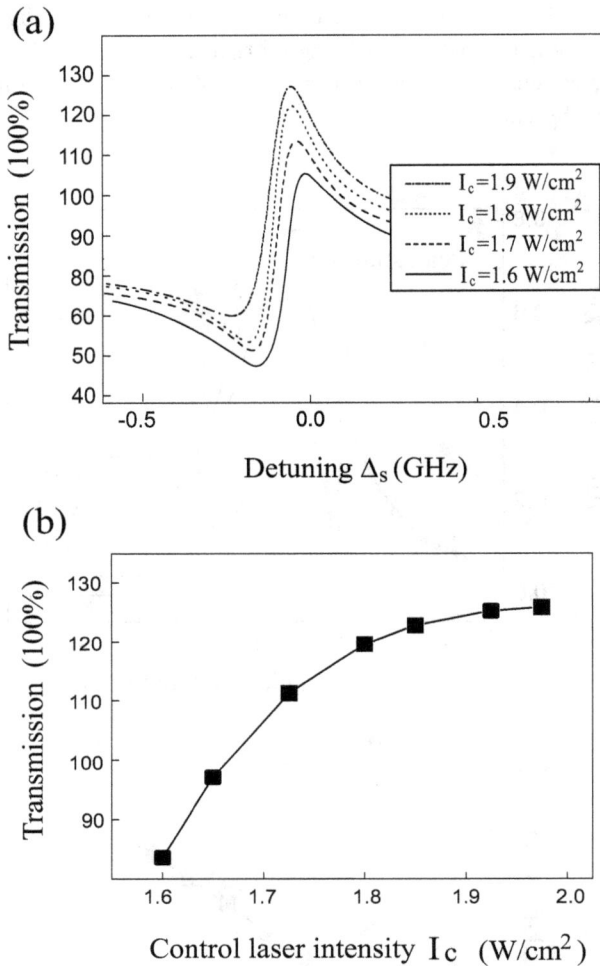

Fig. 10.6   (a) Signal transmission spectrum versus the detuning Δ_s with different control intensities. (b) The characteristic curve of QD-DNA system by plotting the transmission of the signal laser as a function of the control laser intensity. The other parameters used are $\Delta_c = -32\text{GHz}$, $\lambda = 0.8\text{GHz}$, $\omega_D = 32\text{GHz}$, $\tau_D = 3\text{ns}$.

For traditional single optical detection, the fluorescence emission efficiencies of QD-linked biomedicine sensor remain big challenge, according to the copy number of the target, the autofluorescence of background, the coated chemicals, etc. [Zhou (2008)]. However, the emission efficiency can be largely increased in the coupled QD-DNA molecules, driven by two optical fields. Such amplified optical efficiency, different with the traditional detection, is not relevant to the copy number of the target, the autofluorescence of background, and the coated chemicals, which is depends on the quantum interference between the hybrid components and external optical fields. In this case, we anticipate that DNA-linked QD system excited by control-signal technique can be applied to biological imaging. For example, one can inject many QD into the blood of animal, when the quantum dot attaching to abnormal DNA molecules, we first apply a strong control field to the QD, provided by $\Delta_c = -\omega_D$. Then, we apply a second weak signal field across the exciton frequency, the hybrid system can be transparent to the signal field. Thus, the output signal field can be amplified at $\Delta_s = 0$, and the quantum dot can be luminant in optical domain. Furthermore, different vibrational frequencies of DNA molecule and the coupling strengths of QD-DNA lead to different amplitudes of amplification, which are plotted in Fig.10.4 and Fig.10.5. This is the DNA enhanced signal spectroscopy of quantum dot, which will have potential applications in immunoassays, cellular imaging, and clinical diagnosis.

In order to show this DNA enhanced signal spectroscopy more clearly, we further give an example of how to distinct the tumor cells (such as mammary cancer) from normal ones in biomedicine domain when fixing the detuning $\Delta_c = -\omega_D$. Figure 10.7 shows the physical protocol of the discrimination between abnormal cells and normal cells in optical domain. Part (a) of Fig.10.7 is the formation of QD-DNAs system for injecting some quantum dots into biology cell, which including the abnormal DNA molecules and normal DNA molecules. Part (b) gives the conventional biomedicine detection, which is a single optical field detection, such as the fluorescence resonance energy transfer (FRET) process [Zhang (2005)]. The transmission spectrum below part (b) shows that the system attenuates the weak signal beam totally. This dip arises from the usual excitonic absorption resonance. That is why some tumor cells can not emit fluorescence in biomedical assays, even through the attachment of some specific flourescent nanomaterials. However, as we pump on another control laser simultaneously, this attenuation becomes amplification dramatically (Part c). The control beam, like a switch, rapidly controls the transmission of the signal beam. This amplification behavior attributes to the negative part of Fig.10.3(a), which is DNA enhanced spectrum. From the transmission curve of Part (c), we can conclude that DNA-linked QD system can be luminant using this pump-probe technique, which will be a good guide to biomedicine assays. Once the quantum dot is attached to abnormal DNA molecules, we first apply a strong control field at $\Delta_c = -\omega_D$, and then we turn on another weak signal beam across the exciton frequency, in this way, the DNA-linked quantum dot can be transparent and luminant significantly in optical domain by increasing the control laser beam.

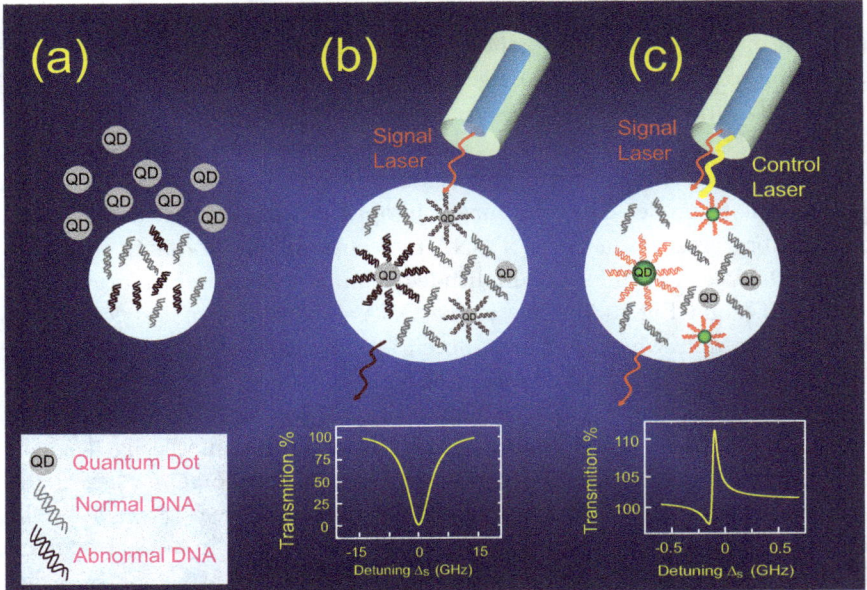

Fig. 10.7 The physical process of the discrimination between tumor cells and normal cells using DNA enhanced spectroscopy in a coupled QD-DNA system. (a) First step, the formation of QD-DNA molecules configuration, while injecting some quantum dots into biology cell. The structure of QD is the same as the Fig.10.1(a). (b) Second step, invisible of such QD-DNA structure in the presence of single optical field. The below curve is the attenuate signal spectrum as a function of detuning between exciton and signal frequency. (c) Third step, luminance of QDs while applying two optical fields. The below plot corresponds to the amplification of signal laser. The other parameters used are $I_c = 1\text{W/cm}^2$, $\Delta_c = -16\text{GHz}$, $\lambda = 0.7\text{GHz}$, $\omega_D = 16\text{GHz}$, $\tau_D = 5\text{ns}$, $\Gamma_2 = 4\text{GHz}$.

Bibliography

T. Liedl, A. Kuzyk, R. Schreiber, Z. Fan, G. Pardatscher, E-M. Roller, A. Högele, F. C. Simmel and A. O. Govorov, DNA-based self-assembly of chiral plasmonic nanostructures with tailored optical response, Nature 483, 311 (2012).

Y. Weizmann, Z. Cheglakov, V. Pavlov and I. Willner, An autonomous fueled machine that replicates catalytic nucleic acid templates for the amplified optical analysis of DNA, Nature Protoc. 1, 554 (2006).

X. Y. Wu, J. Q. Liu, K. N. Haley, J. A. Treadway, J. P. Larson, N. F. Ge, F. Peale, M. P. Bruchez, Immunofluorescent labeling of cancer marker Her2 and other cellular targets with semiconductor quantum dots, Nat. Biotechnol. 21, 41 (2003).

J. Baur, C. Gondran, M. Holzinger, E. Defrancq, H. Perrot, S. Cosnier, Label-free femtomolar detection of target DNA by impedimetric DNA sensor based on poly(pyrrole-nitrilotriacetic acid) film, Anal. Chem. 82, 1066 (2010).

A. Fu, W. W. Gu, C. Larabell, A. P. Alivisatos, Semiconductor nanocrystals for biological imaging, Curr. Opin. Neurobiol. 15, 568 (2005).

M. Hu, Y. He, S. Song, J. Yan, H. T. Lu, L. X. Weng, L. H. Wang, C. Fan, DNA-bridged bioconjugation of fluorescent quantum dots for highly sensitive microfluidic protein chips, Chem. Comm. 46, 6126 (2010).

Z. Zou, D. Du, J. Wang, J. N. Smith, C. Timchalk, Y. Li, Y. Lin, Quantum dot-based immunochromatographic fluorescent biosensor for biomonitoring trichloropyridinol, a biomarker of exposure to chlorpyrifos, Anal. Chem. 82, 5125 (2010).

I. L. Medintz, H. T. Uyeda, E. R. Goldman, H. Mattoussi, Quantum dot bioconjugates for imaging, labelling and sensing, Nature Mater. 4, 435 (2005).

I. L. Medintz, A. R. Clapp, H. Mattoussi, E. R. Goldman, B. Fisher, J. M. Mauro, Self-assembled nanoscale biosensors based on quantum dot FRET donors, Nature Mater. 2, 630 (2003).

K. L. Gilroy, S. A. Cumming, A. R. Pitt, A simple, sensitive and selective quantum-dot-based western blot method for the simultaneous detection of multiple targets from cell lysates, Anal. Bioanal. Chem. 398, 547 (2010).

T. Kim, M. Noh, H. Lee, S.-W. Joo, S. Y. Lee and K. J. Lee, Fluorescence-based detection of point mutation in DNA sequences by CdS quantum dot aggregation, J. Phys. Chem. B 113, 14487 (2009).

X. Michalet, F. F. Pinaud, L. A. Bentolila, J. M. Tsay, S. Doose, J. J. Li, G. Sundaresan, A. M. Wu, S. S. Gambhir, S. Weiss, Quantum dots for live cells, in vivo imaging, and diagnostics, Science 307, 538 (2005).

Y. S. Kim, B. C. Kim, J. H. Lee, J. Kim, M. B. Gu, Specific detection of DNA using quantum dots and magnetic beads for large volume samples, Biotechnol. Bioproc. E 11, 449 (2006).

D. J. Zhou, L. M. Ying, X. Hong, E. A. Hall, C. Abell, and D. Klenerman, A compact functional quantum dot-DNA conjugate: preparation, hybridization, and specific label-free DNA detection, Langmuir 24, 1659 (2008).

C. Y. Zhang, H. C. Yeh, M. T. Kuroki, T. H. Wang, Single-quantum-dot-based DNA nanosensor, Nat. Mater. 4, 826 (2005).

M. Artemyev, E. Ustinovich, I. Nabiev, Efficiency of energy transfer from organic dye molecules to CdSe-ZnS nanocrystals: nanorods versus nanodots, J. Am. Chem. Soc. 131, 8061 (2009).

X. D. Xu, B. Sun, P. R. Berman, D. G. Steel, A. S. Bracker, D. Gammon, L. J. Sham, Coherence optical spectroscopy of a strongly driven quantum dot, Science 317, 929 (2007).

X. D. Xu, B. Sun, E. D. Kim, K. Smirl, P. R. Berman, D. G. Steel, A. S. Bracker, D. Gammon, L. J. Sham, Single charged quantum dot in a strong optical field: absorption, gain, and the ac-Stark effect, Phys. Rev. Lett. 101, 227401 (2008).

L. L. Van Zandt, Resonant microwave absorption by dissolved DNA, Phys. Rev. Lett. 57, 2085 (1986).

B. H. Dorfman, The effects of viscous water on the normal mode vibrations of DNA. Dissert. Abstr. Int. 45, 2213 (1984).

G. S. Edwards, C. C. Davis, J. D. Saffer, M. L. Swicord, Resonant microwave absorption of selected DNA molecules. Phys. Rev. Lett. 53, 1284 (1984).

G. S. Edwards, C. C. Davis, J. D. Saffer, M. L. Swicord, Microwave-field-driven acoustic modes in DNA, Biophys. J. 47, 799 (1985).

C. de Mello Donega, M. Bode, A. Meijerink, Size- and temperature-dependence of exciton lifetimes in CdSe quantum dots, Phys. Rev. B. 74, 085320 (2006).

R. W. Boyd, Nonlinear optics. Amsterdam: Academic Press; p. 313 (2008).

J. Larson, G. Morigi, and M. Lewenstein, Cold Fermi atomic gases in a pumped optical resonator, Phys. Rev. A 78, 023815 (2008).

S. Gupta, K. L. Moore, K. W. Murch, and D. M. Stamper-Kurn, Cavity nonlinear optics at low photon numbers from collective atomic motion, Phys. Rev. Lett. 99, 213601 (2007).

J. F. Marko, E. D. Siggia, Stretching DNA, Macromolecules, 28, 8759 (1995).

C. L. Yuan, H. M. Chen, X. W. Lou, L. A. Archer, DNA bending stiffness on small length scales, Phys. Rev. Lett. 100, 018102 (2008).

R. Gill, I. Willner, I. Shweky, U. Banin, Fluorescence resonance energy transfer in CdSe/ZnS-DNA conjugates: probing hybridization and DNA cleavage, J. Phy. Chem. B 109, 23715 (2005).

B. K. Adair, Vibrational resonances in biological systems at microwave, Biophys. J. 82, 1147 (2002).

N. Amdursky, M. Molotskii, E. Gazit, G. Rosenman, Elementary building blocks of self-assembled peptide nanotubes, J. Am. Chem. Soc. 132, 15632 (2010).

Index